PHILOSOPHY OF COMPUTATIONAL CULTURAL NEUROSCIENCE

This book aims to illuminate theoretical and methodological advances in computational cultural neuroscience and the implications of these advances for philosophy. Philosophical studies in computational cultural neuroscience introduce core considerations such as culture and computation, and the role of scientific and technological progression for the advancement of cultural processes.

The study of how cultural and biological factors shape human behavior has been an important inquiry for centuries, and recent advances in the field of computational cultural neuroscience allow for novel insights into the computational foundations of cultural processes in the structural and functional organization of the nervous system. The author examines the computational foundations of the mind and brain across cultures and investigates the influence of culture on the computational mind and brain. The book explores recent advances in the field, providing novel insights on topics such as artificialism, reconstructionism, and intelligence.

Philosophy of Computational Cultural Neuroscience is fascinating reading for students and academics in the field of neuroscience who wish to take a cultural or philosophical approach to their studies and research.

Joan Y. Chiao is the Director of the International Cultural Neuroscience Consortium. She received her Ph.D. from Harvard University in Psychology and B.S. with Honors from Stanford University in Symbolic Systems.

Essays in Cultural Neuroscience

Series Editor: Joan Y. Chiao

https://www.routledge.com/Essays-in-Cultural-Neuroscience/book-series/ECN

Essays in Cultural Neuroscience showcases scholarly work over a wide range of areas taking a cultural neuroscientific approach. The series is designed to highlight foundations of the fields for interdisciplinary discovery that contribute to our understanding of the neurobiological bases of culture and the mutual construction of culture and neurobiology across evolutionary and developmental timescales. Themes and topics in cultural neuroscience may include a range of cross-disciplinary perspectives, including ethical, scientific, and philosophical issues related to the study of the neurobiological bases of cultural processes. Books are designed to examine the systematic study of theoretical principles, methodological approaches, and empirical paradigms in the fields, including developmental and evolutionary perspectives. This is fascinating reading for students, researchers, and academics across a range of disciplines, including psychology, neuroscience, cultural studies, sociology, and the social sciences more generally.

Also in this series:

Philosophy of Computational Cultural Neuroscience
Joan Y. Chiao

PHILOSOPHY OF COMPUTATIONAL CULTURAL NEUROSCIENCE

Joan Y. Chiao

Routledge
Taylor & Francis Group

NEW YORK AND LONDON

First published 2021
by Routledge
52 Vanderbilt Avenue, New York, NY 10017

and by Routledge
2 Park Square, Milton Park, Abingdon, Oxon, OX14 4RN

Routledge is an imprint of the Taylor & Francis Group, an informa business

Library of Congress Cataloging-in-Publication Data
Names: Chiao, Joan Y., author.
Title: Philosophy of computational cultural neuroscience / Joan Chiao.
Description: New York, NY : Routledge, 2020. |
Includes bibliographical references and index.
Identifiers: LCCN 2020014242 (print) | LCCN 2020014243 (ebook) |
ISBN 9780367347505 (hardback) | ISBN 9780367347512 (paperback) |
ISBN 9780429327674 (ebook)
Subjects: LCSH: Computational neuroscience. |
Cognition and culture. | Philosophy of mind.
Classification: LCC QP355.2 . C465 2020 (print) |
LCC QP355.2 (ebook) | DDC 612.8/233–dc23
LC record available at https://lccn.loc.gov/2020014242
LC ebook record available at https://lccn.loc.gov/2020014243

ISBN: 978-0-367-34750-5 (hbk)
ISBN: 978-0-367-34751-2 (pbk)
ISBN: 978-0-429-32767-4 (ebk)

Typeset in Bembo
by Newgen Publishing UK

CONTENTS

ACKNOWLEDGMENTS

The goal of this book is to introduce philosophical questions that are of core consideration for theoretical, methodological, and empirical approaches in computational cultural neuroscience. The study of how cultural and biological factors shape human behavior has been an important inquiry for centuries. Theoretical, methodological, and empirical foundations in computational cultural neuroscience introduce themes and topics in philosophy at the intersection of areas of the mind and science. Philosophical consideration of the concepts and language of computational cultural neuroscience provides insight into fundamental inquiry of the nature of the mind and the structure of science.

The book is primarily aimed at undergraduate students, graduate students who are studying or faculty who are teaching in philosophy, psychology, neuroscience, or related interdisciplinary courses in computational cultural neuroscience.

The author is grateful to her early mentors for their intellectual interest and support of her work. She is grateful to Nalini Ambady, Jennifer Eberhardt, John Gabrieli, Alexandra Golby, Hazel Markus, and Robert Zajonc of Stanford University, Matthew Lieberman of University of California Los Angeles, Ken Nakayama and Dan Schacter of Harvard University, and Tetsuya Iidaka of Nagoya University for their guidance and encouragement. She is also grateful to Peter Godfrey-Smith, Daphne Kohler, Eric Roberts, Kenneth Taylor, and Tom Wasow of Stanford University for their thoughtful tutelage and encouragement of interdisciplinary studies.

She is with gratitude for her colleagues, Pamela Collins, Beverly Pringle, and Su Yeon Lee-Tauler from the National Institutes of Health for their forbearance and insight. She is also grateful to Tokiko Harada of Hiroshima University, Yoko Mano and Norihiro Sadato of the National Institute for Physiological Sciences, and Jack von Honk of Utrecht University for their intellectual and scientific commitment to international collaboration and international cooperation.

INTRODUCTION

Philosophy of Computational Cultural Neuroscience examines the philosophical foundations of computational cultural neuroscience. Computational cultural neuroscience is a field of study that examines the computational approaches to the study of cultural processes in the structural and functional organization of the nervous system. Theoretical and methodological approaches in computational cultural neuroscience provide the conceptual and empirical tools for understanding the computational principles that guide cultural processes of the brain. Philosophical themes and topics in computational cultural neuroscience inform classic questions in philosophy of mind and philosophy of science.

Theoretical foundations of computational cultural neuroscience consist of computational approaches to elucidate the fundamentals of lawlike regularities and patterns in the cultural processes of neurobiological mechanisms. Theoretical fundamentals of cultural and neurobiological systems rely on natural laws and the lawlike regularity of patterns in natural and artificial systems. Computational approaches to the study of cultural and neurobiological systems consist of tools and instruments for the scientific observation and modeling of cultural processes in neural networks. Cultural and neurobiological systems consist of information-processing mechanisms. Theoretical, methodological, and empirical approaches in computational cultural neuroscience examine how cultural processes are instantiated in neurobiological mechanisms and how neurotechnology can be designed and constructed for the simulation of cultural processes.

Computational approaches to cultural neuroscience allow for the study of the dynamics of information processing in cultural and neurobiological systems. The study of the dynamics of cultural and neurobiological systems consists of the testing of formal models that allow for the characterization and regulation of information-processing mechanisms across multiple levels of analysis. Computational approaches to cultural neuroscience examine how cultural and neurobiological systems represent and regulate the computation of information in biophysical mechanisms across levels of organization of the nervous system.

The design and construction of cultural processes in neurotechnology advance the development and use of applications for the study of cultural processes in neuroscience. The development and use of applications such as the design of cultural programs in computers and devices support the computational model of the cultural brain. The construction of artificial cultural computers and cultural devices that produce cultural computation and cultural content demonstrates the biological plausibility of the cultural brain. The use and application of neurotechnology for the simulation and construction of the cultural level advance an integrative approach to the study of computational cultural neuroscience.

Scientific fundamentals in computational cultural neuroscience address a range of themes and topics in philosophy of mind. The theoretical assumptions of computational cultural neuroscience contribute to themes that address philosophical understanding of the mind as a computing machine at the cultural level. Philosophical considerations of computation in mind and machines at the cultural level provide novel insights into the a range of topics, including the causal-functional role of cultural computation in minds and machines, the role of cultural computation in the production of the cultural level, and the impact of cultural computation for the interaction of minds and machines. The scientific and technological progress from advances in computational cultural neuroscience informs the ethical considerations of cultural neurotechnology for individuals and society.

Philosophical themes from computational cultural neuroscience also advance core considerations in philosophy of science. Research in computational cultural neuroscience informs the impact of scientific tradition and discovery approaches on the nature of explanation. Computational approaches in cultural neuroscience advance the notion that experience and reality is a scientific construction. The use of computational discovery approaches of digital and immersive environments represents the construction of experience and reality in the mind and in the world. The use of technological progress for the construction of experience and reality represents an expansion of mental life from naturalistic into artificial settings. The design and construction of digital and immersive environments for the interaction of minds and machines represents a standard of reform from scientific to virtual realism.

Philosophical Foundations

The goal of this book is to examine the philosophical foundations of computational cultural neuroscience. Themes in computational cultural neuroscience consist of a range of theoretical and methodological foundations in the computational approaches to understand the mutual influence of culture and neurobiology as part of the physical system of possible worlds. The philosophical inquiry into computational principles of cultural computation in the mind and in the world opens a range of core considerations concerning the fundamental nature of the mind and the role of computation in scientific tradition.

Three parts of the book discuss the implications of foundational concepts and processes within the field of computational cultural neuroscience for classic questions of philosophy. Each chapter consists of a rationale that provides a discussion of the philosophical foundations of the scientific approaches and discovery processes in the computational cultural neuroscience field. Themes in philosophy of mind that explore the nature of the mind in the world introduce philosophical stances that serve as a foundation for the standard concepts and processes in computational cultural neuroscience. Theoretical and empirical approaches in computational cultural neuroscience represent an expansion of themes and topics on computational approaches to the study of the mutual influence of culture and neurobiology that inform classic questions in philosophy of science.

In Part I, the first four chapters center on themes in computational cultural neuroscience and their implications for philosophy of mind. Chapter 1 examines the concept of agency, and its functional role as a part of the computation of the mind. The chapter also discusses the concept of cultural agent as a part of the computation at the cultural level of the organized system. Chapter 2 addresses the notion of automatism, as a computational process of the mind. Automatism details the contribution of automatic processing as a computational process at the cultural level of the organized system. Chapter 3 discusses the philosophical implications of the concept of interface for understanding the mind. The chapter expands on the role of interfaces in the interaction of minds and machines. Chapter 4 centers on machine functionalism, or the computational foundations of the mind and its functional role. The chapter details the core considerations in the functional roles of minds and machines of possible worlds.

In Part II, four chapters review the foundational themes in computational cultural neuroscience and their implications for philosophy of mind. Chapter 5 discusses reconstructionism as a theoretical approach to computational modeling in cultural neuroscience. The concept of emergent property as a particular property of the mind reflects a contribution of computational approaches to core concepts in philosophy of mind. The chapter explores the philosophical implications of reconstructionism as a complementary approach to reductionist traditions. Chapter 6 reviews the philosophical stance of machine physicalism, the physical laws and computational principles that comprise patterns and regularities across possible worlds. Machine physicalism entails the physicalist theories of machine computation in the world. Chapter 7 examines computational theory of mind, or computational approaches to understanding mental states. The chapter explores the role of computation in theory of mind. Chapter 8 reviews simulation as a theoretical approach to the computational modeling of the mind. The chapter includes discussion of simulation as part of the computational processes of understanding mental states.

In Part III, the last four chapters consider concepts in computational cultural neuroscience for core considerations of philosophy of science. Chapter 9 addresses artificialism, the contribution of artificial systems to the processes of culture and neurobiology across possible worlds. Artificialism as a philosophical stance explores the role of computation on scientific observation of the natural world, and the impact of computation in the formulation of scientific theory and practice of possible worlds. Chapter 10 includes the theoretical and methodological approaches to machine learning and the philosophical implications of machine learning for explanation. Machine learning entails the philosophical consideration of the role of computational discovery on processes of explanation. Chapter 11 centers on the nature of intelligence and the function of its multiple forms for levels of explanation. Chapter 12 reviews the topic of virtual realism, as a philosophical notion of the impact of computation on the simulation and construction of possible worlds. The chapter explores the nature of scientific theory and observation across possible worlds, and the continuity of scientific and technological progress as precursors to virtual realism.

Philosophical studies in computational cultural neuroscience arise from classical questions in philosophy of mind. Philosophical topics of the mind build from considerations regarding relations of the mind and its realization in the physical world. The notions of the mind and body introduce fundamental elements of the machine in thought experiments for understanding the relational components of the mind and its physical realization. One possible set of mind–body relations is the mind as a machine, reflecting the computational characteristics or possibility of the mind as with functional components of a machine. A second possible set of mind–body relations consists of the brain as a computer, with computational properties of the brain. A third possible set of mind–body relations is the mind and brain as a machine, with lawlike regularities and patterns in the properties of the mind and brain. Different theoretical relations of mind and its physical realization provide novel insight into the computational and functional components of the mind.

The core consideration of the role of computation in cultural processes contributes to a range of questions regarding the nature of the mind. Cultural processes from mental computation refer to the parts and particulars of mental content from cultural computation in the physical world. Mental computation as performance of the computing machine for mental content at the cultural level has multiple physical realizations. The notion of mental computation and its causal-functional role contributes to fundamental inquiry into the nature of the mind across possible worlds.

Classical questions in philosophy of science further inform computational cultural neuroscience. Historical philosophical stances in philosophy of science emphasize the importance of the scientific tradition for understanding the natural world. The expansion of computation as a scientific and technological approach to understanding the world introduces philosophical stances of artificialism that place emphasis on the functional role of artificial systems on scientific theory and practice in possible worlds.

Artificialism comprises the philosophical implications of scientific standards defined through machine computation. Machine computation as a source of computational discovery introduces a causal-functional role for the computing machine in scientific observation and explanation. The role of machine computation in computational discovery opens inquiry into the nature and source of truth theories in scientific traditions. The role of machine computation in the production of informational content to inform explanation presents core considerations regarding the causal network of mental events from a range of sources. The impact of computational discovery on the production of events in the physical world expands philosophical consideration of natural concepts into possible worlds. The expansion of the physical world and its spatiotemporal properties into digital and virtual worlds represents a continuity of scientific and technological progress.

The continuity of scientific and technological progress builds further into core considerations of the nature of experience and reality of possible worlds. The computational discovery of digital and virtual worlds informs the building of evidence-based approaches in the scientific tradition. Computational discovery as digital content acts as an informational source to inform philosophical and scientific standards. The informational content from computational cultural neuroscience informs philosophical and scientific standards in fundamental concepts of the mind and of science.

The role of the scientific tradition in the cultural and public sphere is considerable. The reliance on scientific theory and practice to formulate rationale discourse and to inform societal processes demonstrates the historical importance of the scientific tradition on the public sphere and cultural life. The advancement of scientific and technological progress in our humanistic understanding of culture in mental life stands as an intellectual foundation that informs rational discourse in the public and cultural spheres.

The reliance on technological realism for computational discovery introduces the intention of the mind of the scientist in the simulation and construction of possible worlds. Traditional philosophy of science accounts of scientific observation grapple with the issue of theory-neutral or unbiased observation as a reliable source of information. The application of technology for the simulation and construction of the mind and the restoration and replication of the physical world introduces the intention of the scientific mind as a source of the observation or realism. The simulation and construction of possible worlds place emphasis on the use of scientific instruments in naturalistic forms of scientific observation.

The technological progression of computation as a source of scientific discovery introduces novel considerations regarding the role of scientific instruments in scientific observation and explanation. Early historical events in the progression of the scientific tradition illustrate a fundamental reliance on the human mind for explanatory inference. The contemporary use of technology, computers, and devices for mental content highlights a role of computational discovery in prediction. The modern shift to a technological source of prediction such as from the mind to the computer represents a considerable scientific and technological change. The philosophical implications of computation for cultural and mental life present novel challenges for thoughtful consideration.

Implications

The scientific and technological advancement from research in computational cultural neuroscience informs philosophical inquiry. Research in computational cultural neuroscience integrates theory and methods across the cultural, biological, and computational sciences. Empirical advances in cultural neuroscience demonstrate how culture shapes brain and behavior. Researchers in computational cultural neuroscience aim to demonstrate computational approaches to the study of the cultural brain, particularly the computational principles of cultural processes in the structural and functional organization of the nervous system.

The explorations of computational cultural neuroscience inform core considerations of philosophical inquiry that have numerous implications. In philosophy of mind, the scientific and technological advancement in computational cultural neuroscience introduces concepts and frameworks for understanding the mind at the cultural level. Early historical metaphors of the mind as a machine built into an exploration into the inner workings of the mind as a machine. Traditional notions of the mind as a computing machine presuppose the mind as a machine for cultural computation. The core considerations of the mind as a cultural computing machine preserve a humanistic understanding of the mind in culture.

In philosophy of science, the scientific study of computational cultural neuroscience introduces scientific language and concepts into the structure of science. The scientific language and concepts of computational cultural neuroscience build programs of research and the scientific paradigms of the field. The scientific labor of scientists in computational cultural neuroscience contributes to the social structure of the scientific field. The scientific social structure of the field of study introduces the novel issues of coordination and cumulation in the discovery of evidence-based knowledge in the world.

Future directions of philosophical consideration in computational cultural neuroscience probe the ethical inquiry into the use and application of cultural neurotechnology in science and society. The design and construction of cultural computers and devices advance scientific and technological knowledge of cultural processes of the brain. The use of cultural computers and cultural devices as the tools for the participation and involvement in the cultural sphere opens a range of questions on the significance of computation in cultural development and the advancement of cultural life. The use and application of cultural computation in the public sphere further ethical considerations of the scope and role of cultural computation for individuals and society.

PART I

1
AGENCY

Introduction

Human agency consists of the mental and physical property for social and moral reasoning. At the level of the individual, the mental and physical states of agency in the human mind and brain function as consistent with the social and physical world in a state of truth. Human agency serves as a component of reasoning about the social and physical world that is based on factuality and confirmatory of events of the world that are plausible and existent. Human agency consists of the set of mental and physical properties that are congruent with the social norms and physical laws of the natural world.

The human mind has the capacity to understand the mental and physical states of agency in the other and is known to others through specialized functional properties of the social mind in the world. Agency is perceived as a dimension of perception in the human mind. The detection of agency reflects the perception of a human mind or the functional capacities of the human mind, including self-control, morality, memory, emotion recognition, planning, communication, and thought (Gray, Gray, Wegner, 2007). The perception of human agency in others is highly valued and associated with preference, altruistic motivation, and religiosity.

The perception of human agency in others reflects the detection of another's goal or intention. The inference of agency from social concepts represented in characters (e.g., "robot" or "God") demonstrates components of the mental representation of intentionality or goal state. Social concepts consist of mental trait attributes that are characteristic of other minds and are the content of social inferences of agency in exemplars of a particular type of character. Social concepts may reflect a continuum of agency from detection of minimal intentionality to conscious causal inferences of social attribution.

The functional purpose of mental agency reflects a cultural adaptation. The cultural capacity to understand the mental and physical states of agency in others serves to detect the mental property of other minds that is similar to the human mind. Imagine a possible social world in which all minds consist of the same mental property. The computation to detect the intentionality or goal states of each mind consists of a simple algorithm of an identity relation. Yet, in a possible social world comprised of distinct characters, with distinct mental property, the computation to detect the intentionality or goal state of each mind relies on detection along a gradient or a continuum of the mental property of agency. The inference of other minds as having the mental property of agency is detected not only in humans, but also in supernatural beings. Thus, agency as a mental property guides the mind towards the minds of other beings who are similar, but not necessarily identical, to humans.

The components of mental agency are detected not only in other biological beings, but also supernatural beings. The functionality of mental agency in biological and supernatural beings shares particular properties. The detection of mental agency in biological and supernatural beings may occur automatically and is specialized for the perception of goal and intentional states. The mental states of goals and intentions are the functional property of agency. Mental states of goals and intentions hold representational content that is temporal, based on a future projection of the being in the world. The physical instantiation of the mental states of goals and intentions represents a physical model of future projection generated by the being to navigate the physical world. Physical states of self-prediction, or prediction of the physical and mental state of the self that is ideal or motivated, reflects the internal representations of mental agency in the self.

Physical states of other-prediction, or prediction of the physical and mental states of others' goals or intentions that are ideal or motivated, reflects the detection of the mental agency of other beings. The physical instantiation of other-prediction, or prediction of the physical and mental state of others that is ideal or motivated, reflects the generation of internal representations of other-projection or the other being's self-projection. The physical instantiation of other-projection represents a physical model of future projection generated to predict the navigation of others in the physical world.

The mental and physical states of agency adaptively function to guide social inference and reasoning. Social reasoning of other beings requires an internal model of representational content regarding the future mental and physical states of self and others. The detection of mental agency facilitates the planning and coordination of social thought and action. The conscious deliberation of goals and intentions in self and others enables the long-term planning and coordination of social thought and action. The perseveration of goals and intentions reflects the outcome of planning and coordination of social thought and action in the world.

Mental states of agency are valuable, consisting of truth and falsity. Mental states of goals and intentions that are consistent in the present or can become consistent in the future with mental and physical events in the world comprise agentic mental states that are true or consist of truth value. Mental states of goals and intentions that are unrealized in the present, not idealized or considered unlikely to occur as a mental and physical event in the world, comprise agentic mental states that are false or consist of false value.

The interaction of culture and the environment guides the valuation of agentic mental states through cultural adaptation. Cultural adaptations consist of mental and physical events that have truth value and are factually consistent in the natural world. Cultural adaptations are reinforced as valued patterns of goals and intentions in thought and action. Mental events that are false or inconsistent with the natural world may alter the valuation of agentic mental states and require a change in truth value before reinforcement as a cultural adaptation of a thought or action pattern in the world can occur.

Human agency reflects the goals and intentions of moral thought and action that enhance the genuineness and quality of life in the world. The mental and physical functional properties of the mind and brain have goals and intentions to detect social and physical events that are factually consistent with the natural world. The internal motivation of the agentic mental state is to respond to the sense data in the environment that is true to the intention and goal of the individual. The mental state of agency consists of the goals and intentions that conform to the social norms of the environment and that perform causal reasoning that is forthright to the physical laws of the natural world.

Human agency carries the natural expression of faith and devotion in the religious world. The mental and physical property of faith includes the perception of mind that is consistent and allowable in the religion. The perception of the human mind includes the functional capacities for intention and reasoning of faith. Faithful thought and acts of devotion are a lawlike component of human agency that is characteristic of religion. The internal motivation of the agentic mental state is to respond to the sense data in the environment that is true to the intention and goal of the individual of faith.

Human agency is considered a dimension of the human mind that is related to the perception of moral agency and responsibility. Agency implies that a human can intend or be motivated for moral action, and ascribed responsibility for protection or punishment. Moral attributions, such as those of protection and punishment, are more likely to be related to perceptions of human agency. The detection of human agency occurs as a lawlike regularity of social or moral reasoning of the human mind that occurs in the world, linking the perception of human agency to moral attributions and social explanations of behavior.

Human agency is foundational to the philosophical rationale of human rights. The human capacity for detection and expression of intention in thought and action reflects the plan for lawlike action that is proper and fair in the natural world (Conway, 1894). Human thought as a goal or an intention encompasses a natural and inherent human right for the free and equal expression of lawlike regularity of human mental and physical property. The human capacity for social and moral reasoning, such as attribution of a cause to the social or physical world, reflects the perception of human agency in others. The mental and physical property that allows for the expression and perception of human agency in the self and others is sacred and a natural and inherent right. The consequences of protection and punishment in human agency are as common sense as that in the sacred thought of religion.

Supernatural agency describes characters that represent the lawlike expression of religion. Supernatural agents refer to agents or beings that have goals and intentions that are religious or spiritual, such as gods, goblins, and spirits, and whose actions are counterintuitive to the natural world (Atran & Norenzayan, 2004). Supernatural agents encompass social and moral attributes of benevolence or malevolence, with mental capacities to perform a spectrum of moral or immoral actions that are supernatural. Supernatural agency may constitute the property of benevolent or malevolent characters that exhibit patterns of mental states of "good" or "bad," ranging from plausible and existent to physically implausible or nonexistent in the natural world.

On the one hand, supernatural agents may consist of characteristics that are at least, in part, consistent with social and physical beings of the natural world. Supernatural agents are thought of as models of behavior, characters whose moral actions serve as a model of cultural learning. Supernatural agents that are thought of as a "good" character may encourage cultural transmission of religious beliefs. The moral mental states of the benevolent supernatural agent ("helpful" and "honest") may undergo cultural selection as a set of persistent traits acquired and communicated through social interaction. Supernatural agents that are thought of as a "bad" character may encourage cultural regulation. The immoral mental states of the malevolent supernatural agent ("harmful" and "deceitful") may undergo cultural selection as a set of traits to be detected and regulated for in social interaction.

Beliefs about the physical states of supernatural agents may encompass to some extent characteristics relatable to "good" and "bad" characters in the natural world. Beliefs of the physical states of supernatural agents are thought to include the resemblance of the physical bodies of humans and animals and the physical forms of plants and substances. Religious beliefs of supernatural agents such as gods and goblins refer to physical bodies with configurations of body parts similar to humans and animals. Spirits refer to a type of supernatural agent that may encompass the physical form of a social plant or substance. Supernatural agents may demonstrate "good" characteristics, such as cultural expressions of joy and calm and trust behavior, as well as "bad" characteristics, such as cultural expressions of contempt and anger and deceitful behavior. Beliefs of supernatural agents as possessing physical states of humans and animals consistent with those observed in the natural world engender social reasoning of supernatural and human agents that is intuitive to the natural world.

Beliefs of supernatural agents as possessing physical forms of plants and substances require social reasoning of supernatural and natural agents that is based on the counterintuitive world. The counterintuitive world consists of beliefs that are counterfactual, or inconsistent with fact-based knowledge. Counterfactual beliefs may be physically implausible or nonexistent in the natural world. Counterfactual beliefs contradict facts or factual assumptions of physical and social knowledge of the natural world. Supernatural agents in the physical form of plants and substances consist of mental states and perform actions that contradict social norms and physical laws observed in the natural world. The notion that supernatural agents are omniscient and omnipotent, limitless in knowledge and power contradicts the common-sense knowledge of beings as bounded by knowledge and power of the natural world.

Beliefs of supernatural agents guide perception and action in social inference. Belief of a benevolent supernatural agent may heighten knowledge of moral perception and action, while belief of a malevolent supernatural agent may strengthen motivation for regulation of immoral perception and action. Intuitive beliefs refer to an object or being with a property from the same ontological category (e.g., "melting ice"). Minimally counterintuitive beliefs of supernatural agents include the generation of novel concepts such as an object or being with a property from a different ontological category (e.g., "thinking mineral" and "giggling seaweed"). Maximally counterintuitive beliefs of supernatural agents refer to the generation of novel concepts that associate two properties from a different ontological category with an object or being (e.g., "squinting wilting brick").

The mental capacities and capabilities of supernatural agents may be perceived as primarily representative of the counterintuitive world. Benevolent supernatural agents may be perceived as having supernatural goals and intentions, with capabilities of performing supernatural moral actions. Religious beliefs may imbue benevolent supernatural agents with capabilities to strengthen human protection from threats, deception, and mortality; by contrast, religious beliefs may imbue malevolent supernatural agents with capabilities that weaken human protection.

In the counterintuitive world, the characteristics of supernatural agents may extend beyond the notions of a "good" or "bad" character in the natural world. The beliefs of supernatural agents expand into a reality of beliefs without a truth condition. The capabilities of supernatural agents to protect in the counterintuitive world may necessarily be perceived as physically implausible or inconsistent with the capabilities of humans who perform moral actions in the natural world. In the strictest sense, the criterion of a characteristic of a supernatural agent may be that it is not physically plausible or is in fact nonexistent in the natural world.

Cultural transmission guides social thought and coordination of supernatural agency. Beliefs of supernatural agents arise from sensory pageantry. Sensory pageantry allows for the cultural transmission of religious beliefs and practices about supernatural agents through the coordination of sensory systems for ritual practice. Perception of the supernatural agent occurs across sensory modalities.

Knowledge of the supernatural agent builds from knowledge of the physical world as inferred from sensation.

Cultural transmission of sensory pageantry through social thought and action encourages the transmission of supernatural agency across individuals and groups. The expression of sensory pageantry through language popularizes and spreads the knowledge of religious beliefs and practices. The symbolic representation of sensory pageantry serves as the physical and social realism of supernatural agency in the world. The persistence of religious belief in the mind and the world reflects the potency of cultural transmission and sensory pageantry of supernatural agency.

Agency in Mind and Machine

Agency refers to the mental construct for the detection of agents with distinct goals and internal motivations to reach goals in the physical and social environment. Agency may be detected from contingent movement or interaction between objects and people, such as the movement of dots or geometrical objects that is perceived as dependent on each other or as a singular event (Heider & Simmel, 1944; Blakemore et al., 2003). Mechanical contingency refers to the detection of interaction of objects that occurs due to mechanical causality, such as that characterized in Newton's laws of motion. Intentional contingency refers to the detection of interaction of objects that occurs due to intention or causation at a distance, such as that observed in the social environment.

One characteristic of agency is the detection of animacy or self-propelled movement of an object. An object that moves at a distance through self-propulsion is thought to exhibit animacy or the capacity to produce physical movement through the self. Another characteristic of agency is the detection of nonmechanical contingency or causation at a distance. Causation at a distance refers to an object that follows another object or responds to the movement of another object. Both characteristics of agency comprise fundamental components of theory of mind, or the capacity to understand the mental states of others.

The detection of agency is subserved by brain regions associated with social processes. The superior temporal sulcus (STS) is a brain region located within the upper bank of the temporal lobe consisting of layers of neurons dedicated to the detection of biological motion (Allison, Puce, McCarthy, 2000; Castelli, Happe, Frith, Frith, 2000). The STS is responsive to the perceived and implied movement of body parts important for social communication, including the face, hands, and body. In a neuroimaging study of contingency and animacy, human participants viewed scenes of movement of shapes that varied in contingency and animacy during measurement of neural activity (Blakemore et al., 2003). Neural activity within the left STS was greater when perceiving agency through animacy and contingency of moving shapes. Another neuroimaging study of animacy asked human participants to observe moving shapes in an animate or inanimate background (Wheatley, Milleville, Martin, 2007). Brain regions within the right STS showed greater response when observing animate relative to inanimate moving shapes. The hemispheric lateralization of neural response during the perception of agency across studies may be due to the perception of agency from animacy or contingency. Across studies, neural regions within the STS consist of cellular and molecular mechanisms for the detection of agency from sensory cues of animacy or intentional contingency.

The mental capacity to detect causality and agency from the movement of objects and people arises early during infancy, as a primary component of folk psychology or core knowledge of the physical and social environment. Early knowledge of agency suggests a functional module for theory of mind, such that the sensory input of animacy or intentional contingency necessarily produces an output of agency (Fox & McDaniel, 1982). The detection of agency is considered a fundamental component of theory of mind.

The human capacity to infer agency from biological motion arises early in development. In the first year of life, infants infer causation from biological motion to a greater extent relative to object motion (Saxe, Tenenbaum, Carey, 2005). The movement of the human body leads to causal inferences of agency, an expectation regarding the intentionality or goal state of another mind. The developmental capacity to distinguish physical and social causation during infancy suggests a primary role of agency in building the epistemological foundations of the physical and social world. The ability to infer about the goals and intentions of others represents an early precursor to theory of mind. Mental representations of agency are present from infancy and comprise a foundational component to social inference and reasoning.

The development of supernatural agency or religious belief consists of components of social processing. From the first year of life, infants demonstrate the ability to perceive and detect social information, such as moral reasoning of "good" and "bad" characters (Hamlin, Wynn, Bloom, 2007). The initial state of knowledge of the infant includes the detection of the mental and physical property of beings, including the mental state of moral intention and the physical states of faces and bodies. The perception and detection of others consist of mental representations that relate the mental and physical states of beings as interdependent (Johnson & Wellman, 1982). By childhood, attribution and reasoning of the mental and physical states of others become consciously independent, such that the reasoning about representations of the mental and physical states of others is automatic, but does not necessarily entail each other (Johnson, 1990). During childhood, folk knowledge of the social and physical world builds from initial state reasoning about the properties of other minds and objects.

The neuroscience of religious belief examines the mental and physical representations of supernatural agency in the social and physical world. The neural circuitry of religious belief is associated with a network of brain regions that show functional activation during religious belief. Standard paradigms in the neuroscientific study of religious belief include social processes, such as theory of mind and social reasoning (Han et al., 2008; Kapogiannis et al., 2009). Theoretical models of religious belief center around the mental constructs of supernatural agency and knowledge of spirituality in the social and physical world.

Mental constructs of supernatural agency relate to the perception and knowledge of beings that are moral, such as the concept of God, as well as spiritual notions of gods and goblins. The dimensions of mental experience of religious belief relate to the perception of God's involvement, the perception of God's emotion, and knowledge of religion (Kapogiannis et al., 2009). The dimensions of religious belief include mental and physical knowledge regarding the external reality of a supernatural agent, mental attribution of the supernatural agent, or moral aspects of religion. The knowledge of religion includes the external reality of a supernatural agent and the existence of the social and physical world of the supernatural agent (e.g., "There is a God," "People go to heaven"). Mental attribution of the supernatural agent refers to the qualities of emotional and cognitive mental states of God (e.g., "God is loving," "God protects"). Knowledge of theological and moral aspects of religion relate to doctrinal (e.g., "A source of creation exists") and experiential religious belief (e.g., "Religion provides moral guidance").

The dimension of religious belief related to the perception of God's level of involvement consists of mental and physical property. The mental and physical property of God's level of involvement includes two neural networks of anterior and posterior brain regions within the right hemisphere, a lateral and medial neural network. The lateral neural network of God's level of involvement consists of several brain regions within the right inferior frontal gyrus, the middle occipital gyrus, the middle temporal gyrus, and the inferior temporal gyrus. The medial neural network of God's level of involvement includes the superior medial frontal gyrus, precuneus, and the left inferior frontal gyrus. The functional neural activity within the lateral and medial neural network relates to the perception of a lack of involvement of God. The perception of the presence of God's involvement was not related to a pattern of neural activation within the lateral and medial neural networks of the right hemisphere. These findings suggest that neural networks are recruited when there is a lack of detection of God's involvement, suggesting greater enhancement of neural resources for the processing of information in the environment. Neural networks associated with theory of mind may be recruited to a greater extent due to the increased cognitive load afforded to the perception of agency, in the absence of the perception of supernatural agency. Thus, the mental and physical property of religious belief consists of the increased recruitment of brain regions for theory of mind during the perceived absence of God's involvement.

The dimension of religious belief associated with the mental attribution of God consists of mental and physical property. The perception of the positive emotional states of God, such as the perception of God's love, is associated with neural response within the right middle frontal gyrus, while the perception of the negative emotional states of God is related to neural activity within the left middle temporal gyrus. The perception of God's emotional states engages brain regions involved with the detection and regulation of emotional theory of mind. The mental and physical property of religious belief related to the mental attribution of supernatural agency, such as God, is localized to specific brain regions involved in the detection and regulation of emotional theory of mind.

The dimension of religious belief consisting of religious knowledge includes mental and physical property. The mental property of doctrinal religious knowledge refers to abstract beliefs of theology, while those of experiential religious knowledge refer to abstract beliefs of morality in society. Physical property of doctrinal religious knowledge consists of a network of brain regions related to abstract religious beliefs. The neural network of doctrinal religious belief includes the right inferior temporal gyrus, right middle temporal gyrus, right inferior parietal gyrus, left cingulate gyrus, and left superior temporal gyrus. Brain regions within the neural network of doctrinal religious belief are associated with theory of mind, including agency detection. Doctrinal religious belief relies on a neural network of brain regions for the detection of intention and action of agents.

The neural network of experiential religious belief consists of the bilateral calcarine gyrus, left fusiform gyrus, left precuneus, left precentral gyrus, and left inferior frontal gyrus. Brain regions within the neural network of experiential religious belief are related to social perception and communication. Mental and physical representations of social knowledge are recruited to a greater extent when moral reasoning about experiential religious belief. Experiential religious belief refers to the moral reasoning within society and the mental and neural representations of social processes, such as the mental attributions of people.

Mental state attribution of supernatural agency engages brain regions specialized for social knowledge. In a neuroimaging study of religion and the self, religious and nonreligious Christians performed a character trait attribution task (Han et al., 2008). Participants encoded character traits during social evaluation of themselves or a public person. Findings from the study show that neural activity within the ventral portion of the medial prefrontal cortex (VMPFC) increases for self relative to public person evaluation for nonreligious relative to religious participants. For religious Christians, neural response within the dorsal portion of the medial prefrontal cortex (DMPFC) is greater for self relative to public person evaluation, but not for nonreligious participants. Magnitude of functional activity within the DMPFC was significantly related to the magnitude of importance of Jesus' judgment in social evaluation. For people with religious belief, thinking about the character of one's self and others is related to neural processing for social inferences of the intention and beliefs of others. Attribution of mental states for people with religious belief is associated with enhanced activity in brain regions important for social inference about supernatural agency. The functional organization of brain activity shows differentiation during the social evaluation of religious and nonreligious belief.

The influence of religious belief on behavior is regulated by genes. Gene-by-environment interactions posits that the interaction of genes with environment regulates behavioral expression. Heightened exposure to religion in the environment interacts with genes to regulate levels of prosocial behavior (Sasaki et al., 2013). Greater exposure to religion in the social environment enhances prosocial behavior or willingness to help others for people with genetic sensitivity. People who carry the two- or seven-repeat allele of the dopamine receptor gene (DRD4) show greater prosocial behavior when temporarily exposed to religion. Genetic sensitivity for the expression of moral behavior is heightened when exposed to religion. These findings show that genetic sensitivity enhances the influence of religion for prosocial behavior.

Genes regulate the efficiency of neurotransmission underlying complex behavior. The influence of religion on genetic sensitivity for prosocial behavior suggests that mental and neural representations for morality are similarly strengthened with exposure to religion. People who are genetically sensitive to religion may show heightened mental and neural architecture for moral expression. The moral expression of behavior reinforces mental and neural processes underlying the behavioral regulation of religion.

Culture and Agency

Coevolutionary theories of agency posit that cultural and genetic inheritance regulate mental and neural mechanisms of agency. Through exposure to environmental demands, organisms undergo selection of characteristic traits that produce adaptive behavior. Coevolutionary processes show that organisms not only undergo genetic selection in response to environmental demands, but also shape their environment through processes of cultural niche construction. In processes of cultural selection, organisms create and build environmental niches that strengthen behavioral adaptations.

Based on coevolutionary theory, mental and physical representations of agency are reinforced through processes of cultural and genetic selection. Cultural traits that build mental and neural mechanisms of agency, such as theory of mind and social reasoning, may undergo reinforcement through genetic selection. Genetic variation of specific genes associated with social processes is enhanced as the expression of agency functions as a cultural adaptation.

Human agency as mental and physical property of culture consists of the patterns of goals and intentions that are representative of the group. The detection and expression of human agency of the cultural group consist of the distinct sets of characteristic goals and intentions that function as cultural adaptations. For instance, the mental and physical representations of agency in an individualistic culture include the pattern of mental and neural representations that comprise a lawlike regularity for majority, while that of a collectivistic culture includes the pattern of representations that consist of a lawlike regularity for uniformity in goals and intentions and function as social norms of the environment.

The regularity in goals and intentions of a culture acts as a cultural adaptation that may be reinforced through genetic selection. Cultural regularity in goals and intentions may similarly be strengthened not only through reproduction, but also through cultural niche construction. Cultural changes built into the environment can magnify patterns of cultural adaptation in the mind and brain. The mutual influence of cultural and genetic selection on mental and physical representations of agency comprises a multilevel mechanism of social processes.

The cultural regulation of agency is represented in the mind and brain through multilevel mechanisms. Multilevel mechanisms of the mind and brain consist of three levels of analysis: computational, algorithm, and physical implementation (Marr, 1982). At the computational level, the set of mental states comprising goals and intentions that are congruent with one's culture reflects a component of tasks that support the computational reinforcement of cultural norms. At the algorithm level, the neural mechanism of agency is conceptualized as formal procedure with a given input to produce a specific output. An algorithm for cultural regulation of agency may consist of neural mechanisms that transform the sense data of cultural input from the world into physical representations of goals and intentions in the mind and brain. At the level of physical implementation, the set of neural states that comprise cultural goals and intentions refers to the physical implementation of the cultural agency within neural organization.

The cultural regulation of agency may rely on conceptual models of goal and intentional states. Models of cultural regulation consist of a set of directional relations of goal states and cultural norms, practices, and beliefs. Processes of cultural regulation include the conceptual models of goal and intentional states that contribute to the relations between heritage and host culture. Cultural regulation relies on processes of acculturation.

The acculturation process consists of maintenance of social relations among heritage and host culture. The content of goals and intentions affects the strength of relations among the heritage and host culture. Cultural agency, or orientation towards culture-specific goals and intentional states, affects how people acculturate from heritage to host culture. Intercultural processes are affected by lay theories of race (Hong, Chao, No, 2009). People who are incremental theorists hold a dynamic constructionist orientation and are more likely to perceive ability as based on effort during difficulty, while people who are entity theorists maintain an essentialist orientation and are more likely to perceive ability as immutable essences (Dweck, 1986). Incremental theories of culture refer to a learning orientation to the goals and intentional states of the heritage and host culture. People with an incremental theory of race may perceive the acquisition of distinct cultural orientations as effort-based and strengthen goal and intentional states of cultural integration or assimilation, orientations that benefit the heritage and host culture. On the other hand, entity theories of race may rely on an essentialist orientation and pursue goals and intentional states of cultural marginalization or exclusion, orientations that primarily benefit the heritage culture. Thus, lay theories of race contribute to the set of component tasks that compute goals and intentions as processes of cultural agency.

Developmental processes illustrate the transmission of cultural representations of agency through social interaction. During childhood, social interactions between parent and child guide the cultural construction of self-concept and identity (Wang, 2006). Cultural views of the self guide parents and children to emphasize distinct aspects of culture-specific knowledge and understanding of the social and physical world. For instance, European-American children are more likely to emphasize personal aspects of the self, such as character traits, while Chinese children are more likely to refer to social aspects of the self, such as situation-bound characteristics. European-American parents strengthen memory construction with conversational recollection with children that consists of related memory events, while Chinese parents show a greater inhibitory role in providing feedback that strengthens memory construction, preferring to emphasize the situational importance of memorable events. Through the cultural construction of self-concept, parents and children share internal representations of autobiographical intentions and goals that are culture-specific. Cultural variation in self-concept arises early in childhood through the mutual construction of internal representations of self-concept and social reasoning between parents and children.

Dynamic constructivist approaches to culture show how cultural change can occur through minimal experience with situational sense data. Cultural priming demonstrates the influence of culture on mental and representations of social processes (Hong, Morris, Chiu, Benet-Martinez, 2000; Oyserman, Coon, Kemmelmeier, 2002). Cultural orientations of individualism and collectivism lead to distinct sets of social processes and behaviors. Temporarily heightening awareness of cultural orientation leads to stronger internal representations of cultural information from sensory input. Dynamic constructivist approaches to culture suggest that internal representations of culture can perform transformations and hold interpretations of sense data dependent on the situation.

Multicultural minds build and maintain internal representations of distinct cultural orientations and demonstrate the capacity to switch from one cultural frame to another based on the environment. Cultural priming of individualism enhances mental and neural representations of general self processes, while priming of collectivism strengthens representations of contextual self processes. Cultural priming modulates patterns of neural activation within brain regions specialized for self processing. Cultural priming of individualism enhances neural activation within social brain regions during general self processing, while those of collectivism strengthen during contextual self processing.

The process of switching from one cultural orientation to another demonstrates the dynamic and flexible characteristic of cultural networks. Neural networks within the prefrontal cortex show distinct patterns of neural activation dependent on culture-specific sense data (Harada, Li, Chiao, 2010). The detection of sensory input that is consistent with the cultural orientation strengthens information-processing mechanisms that maintain representations of culture. During cultural priming, multilevel mechanisms strengthen the pattern of activation for one cultural orientation to a greater extent relative to another. The relative strength of the representations within the cultural network may similarly weaken, dependent on processes of cultural priming.

Computational modeling of culture characterizes the dynamical basis of fundamental cultural processes. At the neural level, cultural orientation may exert a top-down influence on processing of control and biasing of the interpretation of sense data; cultural orientation may similarly demonstrate a bottom-up influence on information processing through transformations of sense data. Dynamics of cultural networks show the spatiotemporal characteristics of information-processing mechanisms and their causal interactions. The act of switching from one cultural orientation to another suggests computational models of culture comprised of multistage processing and dynamic causal interactions between network nodes. Genetic variation in cultural processes suggests that intrinsic properties of the nervous system show information-processing mechanisms specific to culture. Computational modeling of cultural developmental processes refers to the characterization of social dynamics within the cultural network. Modeling of the social dynamics of the cultural network consists of positing the causal interactions of network nodes across multilevel mechanisms. The structural and functional properties of the nervous system provide multiple levels of organization for the maintenance and generation of cultural adaptation.

Cultural Agency

Cultural agency comprises the mental and physical property for social and moral reasoning produced through processes of cultural niche construction. Through interactions with the environment, the organism changes and builds the cultural niche as a byproduct of adaptation. Cultural agency refers to the property of the organism that strengthens the cultural niche through social and moral reasoning. The property of cultural agency may consist of mental and physical states that are natural or artificial. Natural cultural agency includes the inherent or intrinsic mental and physical property for social and moral reasoning. Artificial cultural agency consists of the man-made mental and physical property for social and moral reasoning.

Cultural agency refers to a type of character with culturally normative properties that can be materialized in human and environment interaction. The notion of a cultural agent reflects the capability of a nonliving entity to represent a set of the mental and physical properties of a living being, particularly the mental properties related to folk biological concepts of "animate" and "intentionality." The cultural agent as a nonliving entity may encompass specific characteristics such as fictional, computer-generated, and cartoonlike, acting as a physical interface between the cultural agent and the environment in a social interaction.

The cultural agent communicates in a social interaction as a character in a manner congruent with cultural norms. The verbal and nonverbal communication of the cultural agent may reflect a cultural system that is artificial, man-made, or computer-generated. In an artificial cultural system, cultural values, practices, and beliefs maintain and reinforce specific functional routines of verbal and non-verbal communication between the human and the computer. An artificial cultural system may consist of values, practices, and beliefs that maintain and reinforce specific functional routines of communication that are normative of the epistemology of the human or the computer.

The cultural agent embodies and communicates natural or artificial cultural intelligence. The cultural agent consists of mental and physical knowledge of cultural systems. Mental and physical knowledge of cultural systems may be generated and maintained in a natural or artificial world. Natural cultural intelligence refers to the mental and physical knowledge of cultural systems that occurs from learning in the natural world, such as that afforded by ecological and intellectual inheritance. Natural cultural intelligence may include the epistemology acquired through classic vertical or horizontal transmission, such as cultural knowledge acquired from parents to offspring or from teachers to students.

Artificial cultural intelligence refers to the mental and physical knowledge of cultural systems that occurs from learning in a machine world, such as knowledge generation from machine learning. Artificial cultural intelligence encompasses mental and physical knowledge of cultural systems that are generated and maintained in the machine world. Artificial cultural intelligence may be generated through machine learning in an automated fashion. Artificial cultural intelligence may generate knowledge of cultural systems that reflects the spatiotemporal properties of the machine world. The spatiotemporal properties of the machine world may be distinct from that of the natural world in several ways. The machine world may refer to man-made properties of a physical world. The machine world may consist of the properties of a world in a computer, the man-made properties of a mental world, or the man-made properties of a physical world.

Cultural Agency as Cultural Computation

The cultural agent has the capability to interact with the human in a manner consistent with cultural norms. The cultural agent shows cultural competency in social interaction and is able to perform cultural practices, and to embody cultural values and beliefs. Cultural agents that are designed and trained for social interaction with humans show computational properties of the mind that reflect the epistemology of human culture.

Computational properties of the mind include component tasks and algorithms that comprise the mental capacities for cultural processes. Computational properties of the cultural mind can be decomposed into component tasks, such as affiliation and attachment, social communication, and the perception and understanding of self and others at the cultural level. Mental algorithms of cultural processes reflect functional procedures that produce a specialized output given a specific input. Functional modules for social processes refer to procedures that are specialized to produce a social output given a specific sensory input at the cultural level. Human culture is an adaptive system responsive to environmental and ecological pressures.

Individualistic cultural agents show computational properties of self-expression, autonomy, and uniqueness. Individualistic cultural agents display verbal and nonverbal communication that is socially and emotionally expressive, reinforcing the sense of autonomy and uniqueness of the cultural agent. Collectivistic cultural agents show computational properties of social harmony, hierarchy, and conformity. Collectivistic cultural agents demonstrate verbal and nonverbal communication that emphasizes social harmony, recognizes hierarchical social displays, and observes the social norms of the interaction.

Supernatural agents as a type of cultural agent are characterized by particular computational properties in cultural computation. Supernatural agents as omniscient refer to minds that contain computational properties that are unquantifiable and unconstrained. Supernatural agents as omniscient represent infinite knowledge as infinite in the computational ability to generate and recognize knowledge. Supernatural agents as omnipotent refer to minds with the unconstrained computational ability, that is, the ability to compute in an unlimited or unbounded manner.

In human–computer interaction, the cultural agent acts as a computational interface. The cultural agent enhances computation in the social interaction of a human and a computer in a number of ways. The cultural agent provides the sensory input, such as facial and nonfacial communication, for the human to decode and encode. The computer generates the nonverbal communication of the cultural agent to express to the human. The computer may detect and decode social communication cues of the human and dynamically transform the social communication of the cultural agent to respond to the inferred mental states of the human. The verbal and nonverbal communication between the human and the computer is computational at the phase of computer-generated social and emotional perception and expression.

Conclusion

The perception of the cultural agent is computationally demanding, requiring the coding and decoding of communicative signals that are byproducts of cumulative cultural evolution. Functional architecture in the mind and brain that has evolved to understand the mental states of others in the natural world may require specialized mental and neural resources to represent and interpret the communicative signals of the cultural agent in the machine world. For the computer, the perception of the human is computationally demanding, requiring the coding and decoding of communicative signals that are byproducts of biological evolution. The capability to represent and respond to the communicative signals of the human at the cultural level requires the functional capacities of computation in minds and machines.

References

Allison, T., Puce, A., McCarthy, G. (2000). Social perception from visual cues: role of the STS region. *Trends in Cognitive Science, 4*, 267–278.

Atran, S. & Norenzayan, A. (2004). Religion's evolutionary landscape: counterintuition, commitment, compassion, communion. *Behavioral and Brain Sciences, 27(6)*, 713–730.

Blakemore, S.J., Boyer, P., Pachoy-Clouard, M., Meltzoff, A., Segebarth, C., Decety, J. (2003). The detection of contingency and animacy from simple animations in the human brain. *Cerebral Cortex, 13(8)*, 837–844.

Castelli, F., Happe, F., Frith, U., Frith, C. (2000). Movement and mind: a functional imaging study of perception and interpretation of complex intentional movement patterns. *Neuroimage, 12*, 314–325.

Dweck, C. (1986). Motivational processes affect learning. *American Psychologist, 41(1)*, 1040–1048.

Fox, R. & McDaniel, C. (1982). The perception of biological motion by human infants. *Science, 218(4571)*, 486–487.

Gray, H.M., Gray, K., Wegner, D.M. (2007). Dimensions of mind perception. *Science, 315*, 619.

Hamlin, J.K., Wynn, K., Bloom, P. (2007). Social evaluation by preverbal infants. *Nature, 450(7169)*, 557–559.

Han, S., Mao , L., Gu, X., Zhu, Y., Ge, J., Ma, Y. (2008). Neural consequences of religious belief on self-referential processing. *Social Neuroscience, 3(1)*, 1–15.

Harada, T., Li, Z., Chiao, J.Y. (2010). Differential dorsal and ventral medial prefrontal representations of the implicit self modulated by individualism and collectivism: an fMRI study. *Social Neuroscience, 5(3)*, 257–271.

Heider, F. & Simmel, M. (1944). An experimental study of apparent behavior. *The American Journal of Psychology, 57*, 243–259.

Hong, Y.Y., Chao, M.M., No, S. (2009). Dynamic interracial/intercultural processes: the role of lay theories of race. *Journal of Personality, 77(5)*, 1283–1309.

Hong, Y.Y., Morris, M.W., Chiu, C.Y., Benet-Martinez, V. (2000). Multicultural minds: a dynamic constructivist approach to culture and cognition. *American Psychologist, 55(7)*, 709–720.

Johnson, C.N. (1990). If you had my brain, where would I be? Children's understanding of the brain and identity. *Child Development, 61(4)*, 962–972.

Johnson, C.N., Wellman, H.M. (1982). Children's developing conceptions of the mind and brain. *Child Development, 53(1)*, 222–234.

Kapogiannis, D., Barbey, A.K., Su, M., Zamboni, G., Krueger, F., Grafman, J. (2009). Cognitive and neural foundations of religious belief. *Proceedings of the National Academy of Sciences, 106(12)*, 4876–4881.

Marr, D. (1982). *Vision*. New York: Freeman.

Sasaki, J.Y., Kim, H.S., Mojaverian, T., Kelley, L.D., Park, I.Y., Janusonis, S. (2013). Religion priming differentially increases prosocial behavior among variants of the dopamine D4 receptor (*DRD4*) gene. *Social Cognitive and Affective Neuroscience, 8(2)*, 209–215.

Saxe, R., Tenenbaum , J.B., Carey, S. (2005). Secret agents: inferences about hidden causes by 10- and 12-month-old infants. *Psychological Science, 16(12)*, 995–1001.

Wang, Q. (2006). Culture and the development of self-knowledge. *Current Directions in Psychological Science, 15(4)*, 182–187.

Wheatley, T., Milleville, S.C., Martin, A. (2007). Understanding animate agents: distinct roles for the social network and mirror system. *Psychological Science, 18(6)*, 469–474.

Further reading

Blakemore, S.-J. & Decety, J. (2001). From the perception of action to the understanding of intention. *Nature Reviews Neuroscience, 2*, 561–567.

Cassell, J., Sullivan, J., Prevost, S., Churchill, E.F. (2000). *Embodied conversational agents*. Cambridge: MIT Press.

Conway, M.D. (Ed.) (1894). The writings of Thomas Paine (4 vols). New York: G.P. Putnam's Sons.

Deheane-Lambertz, G., Deheane, S., Hertz-Pannier, L. (2002). Functional neuroimaging of speech perception in infants. *Science, 298(5600)*, 2013–2015.

Frith, C.D. & Frith, U. (2007). Social cognition in humans. *Current Biology, 17*, R724–R732.

Gallagher, H.L., Happe, F., Brunswick, N., Fletcher, P.C., Frith, U., Frith, C.D. (2000). Reading the mind in cartoons and stories: an fMRI study of 'theory of mind' in verbal and nonverbal tasks. *Neuropsychologia, 38*, 11–21.

Leslie, A. & Keeble, S. (1987). Do six-month old infants perceive causality? *Cognition, 25*, 265–288.

Newton, I. (1687). *Philosophiae naturalis principia mathematica*. London.

Oakes, L.M. & Cohen, L.B. (1990). Infant perception of a causal event. *Cognitive Development, 5*, 193–207.

Ruby, P. & Decety, J. (2001). Effect of subjective perspective taking during simulation of action: a PET investigation of agency. *Nature Neuroscience, 495*, 546–550.

Tremoulet, P.D. & Feldman, J. (2000). Perception of animacy from the motion of a single object. *Perception, 29*, 943–951.

Vogeley, K. & Roepstorff, A. (2009). Contextualising culture and social cognition. *Trends in Cognitive Science, 13(12)*, 511–516.

2
AUTOMATISM

Introduction

Philosophical notions of the mind consider the mental and physical properties across possible worlds. The properties of the physical world as a physical system are consistent. Physical properties of the physical world show a range of mechanisms and their parts. The mind as a biological computing machine consists of physical properties that perform causal-functional roles. The physical properties of mental and machine computation demonstrate the automatic production of specific input–output relations based on formal procedure.

The biological computing mind performs mental functions that constitute automatic processing. The biophysical mechanisms of mental computation consist of neural mechanisms that perform automatic processing of information. The automatic processing of information is a mental property based on the functional specialization of modules of the mind. The automaticity of information processing is one of the most basic properties of mental computation. In the biological computing machine, automatic information-processing mechanisms describe the lower-level processing of sense data based on the simple association of a stimulus with a response.

The automation of organized systems describes the automated production of machine computation and its physical properties. Automatism as a property of real systems describes the production of specific input–output relations from machine computation. Probabilistic automatons as computing machines produce specific machine output from physical state transitions that are preset based on instructions. In artificial living systems, cellular automatons demonstrate patterns of self-organization from the interaction of local elements. Artificial life shows the basic feature of self-replication as an automated pattern in the artificial living system. Computational models of living systems describe the automated production of basic features of living systems that arise as global patterns. The automation of information production from machine computation is consistent with the earliest notions of the functional role of the computing machine.

Automatism

Automatism is related to the set of things in the world that govern automatic real systems. Automatic real systems demonstrate properties of control and perform actions without conscious thought or volition. Automatic real systems demonstrate characteristics of physical laws. That is, the automaticity of the thing in the world of the real system is governed through physical law. Physical laws describe real systems and observe several characteristics. A physical law is a generalization that is true and applies to all of space and time. A law of nature describes how things have to be or necessarily how things are in the world. A scientific law describes how things are without exception. A scientific law can also refer to regularity that occurs within a historical context or a pattern in nature that holds with regularity.

In a computational context, automatism may refer to a set of things called automatons. The automaton is a set of things that have the properties of automaticity and can move or act by themselves. A thing may be considered an automaton if it can move or act by itself in a mechanical sense or based on instructions previously given to it through a computer. The automaton may refer to a thing preset with encoded instructions to automatically perform specific actions.

The set of things in the physical world that exhibit properties of autonomous agents demonstrate automatic information processing. Autonomous agents as biological organisms perform automatic functions. The human mind as an autonomous embodied agent is comprised of automatic information-processing mechanisms for the mental performance of basic functions. The mental performance of automatic information processing consists of the basic mental functions that facilitate adaptive response.

The study of automatic processing refers to basic neural mechanisms of functional specialization in the structural organization and function of the nervous system. The mind and brain refer to the set of mental and physical states that observe properties of automaticity. Automaticity in the mind and brain may refer to the set of mental and physical states that occur without conscious thought or volition. Mental states that are unconscious or reflexive thought processes are considered automatic. Physical states that occur in the organization of the nervous system and are associated with unconscious or reflexive thought processes are considered automatic. Mental and physical states in the mind and brain that are modular or are specialized to perform a particular function are thought to automatically occur. Automaticity in the mind and brain may also refer to the set of mental and physical states that perform functions of self-regulation.

The organization of the nervous system consists of multiple neural systems that demonstrate functional specialization of the mind. The functional specialization of the mind as modules entails information processing that is automatic, rapid, innate, encapsulated, and physically instantiated (Fodor, 1983). The automatic processing of specific tasks refers to the production of thought that is lower-level. Mental content as modules of automatic processing implies basic processes of thought. Automatic mental processing consists of information-processing mechanisms that produce direct responses based on autoassociation.

Automatic and controlled processing reflect a computational principle of the structural and functional organization of the nervous system. Automatic neural circuitry consists of neural information-processing mechanisms for performing computational algorithms in an automated fashion. Given specific input, automated mechanisms perform a specialized operation to produce accurate output. The cognitive theory of functional modularity posits that automated processing mechanisms within the mind and brain perform specialized functional operations to produce accurate output. Simple cognitive operations, such as face and place recognition, occur rapidly and with minimal effort in the presence of specific input, suggesting that the automated mechanisms for performing such cognitive operations are necessary and sufficient as patterns of cognitive processes.

Controlled processing refers to the regulation of automaticity with effort. Controlled neural circuitry consists of neural circuitry for down-regulation and up-regulation of automatic mechanisms. Controlled processing refers to information processing that is considered higher-level. Controlled processing facilitates the flexible adaptation of thought to the environment. The use of controlled information-processing mechanisms demonstrates the regulation of automatic thought.

Cultural Automatism

Cultural automatism is the set of things in the world that show properties for the automatic production of cultural processes in the physical world. Cultural automatism includes cultural processes and the mental and physical properties that occur in an automatic manner in the physical world. Cultural automatism refers to the set of things in the physical world called cultural automatons. The cultural automaton is a set of things that have properties of automaticity and can move or act by themselves in the cultural system and in the physical world.

A thing may be considered a cultural automaton if it can move or act by itself in a mechanical sense as part of the cultural system in the physical world. The cultural automaton is a thing with preset instructions that are encoded or that automatically performs specific actions with instructions previously given to it by the computer in the cultural system. The cultural automaton may automatically detect all of the possible input in the environment that is a component part of its cultural system. Similarly, the cultural automaton may compute all of the possible output in the environment that is producible within its cultural system. Cultural automatons may coordinate their movement or action to create and maintain their cultural system in the physical world (e.g., rover; outer-space objects). The cultural automaton may detect as input automatic cultural property in the superenvironment.

Cultural automatons in the superenvironment can automatically detect and receive novel cultural property, including cultural events, objects, and organisms. The cultural property of the superenvironment comprises component parts as preset output with instructions to produce specific action. Novel cultural property is a mechanism of cultural maintenance in the superenvironment. The cultural automaton automatically produces novel cultural property that is consistent with the superenvironment. Maintenance and production of novel cultural property in the superenvironment ensure the consistency of programs of action and sustainable levels of output.

Cultural traditions are maintained through the participation and production of cultural practices in groups of people defined by shared geography, ancestry, language, customs, and heritage. Cultural traditions ensure the participation in and production of cultural practices with lawlike regularity. Cultural traditions share the automatic patterns of cultural thought reflected in the values and beliefs of the cultural system. Cultural traditions in the superenvironment are a far-reaching and goodly mechanism of cultural maintenance.

Cultural traditions are strengthened through the automatic patterns of cultural thought. Cultural systems of thought consist of sets of automatic social thoughts of enjoyment. Social thought comprises the mental property of objects and their attributes. Mental property of social thought consists of the set of social mental states. Social mental states occur as mental events characterized in a range of social phenomena from social sense data to mental state attribution. Social mental states of objects and their attributes can be acquired with emotional valence. Social mental states with emotional valence build the set of mental events that include mental state attributes and other objects associated with positive and negative emotional valence. Social mental events are a type of automatic social thinking that builds patterns of cultural thought of enjoyment.

Cultural change in the property of the superenvironment reflects the use of technological innovation for production of encoded instructions to change specific action through preset output. Levels of cultural property in the superenvironment reflect trades of programs of action and preset output. Optimal levels of trade with cultural property are consistent with a maximal possible output of the superenvironment. Novel cultural property allows for the assembly of novel preset output to produce novel programs of action. Cultural property produced in the social, economic, and political sphere reflects the implementation of programs of action.

The programs of action and output of cultural automatism in the superenvironment show the capability for protection and empowerment that governs automatic real systems. Cultural maintenance of the superenvironment provides protection of resources and services that reflect an advancement in quality of life. Cultural participation in the superenvironment ensures the protection inherent of regard and rank. Cultural automatons can provide access to resources and services that serve as basic protections. Novel cultural property as input into the superenvironment can assist to coordinate as the output the development and implementation of programs of action for individuals and groups. Novel cultural property contributes to the advancement of cultural development of nations, societies, and individuals.

The production of cultural change in the superenvironment can improve access to resources and protections. High levels of resources and protections in the superenvironment reflect a preparedness to show response and recovery to threats. Cultural automatons contribute to a high level of protection in the superenvironment through the production of programs for specific actions and output. Cultural changes in levels of protection manifest as automatic upward transformations of cultural property in the superenvironment to produce high levels of preparedness and low levels of threat. The maintenance and production of cultural property are beneficial for high levels of protection in the superenvironment.

Cultural changes in the superenvironment can alter levels of protection through the automatic transformation of the truth value of cultural property in the physical world. The maintenance of cultural property in a state of its maximal truth value in the physical world may produce automatic upward transformation for its protection. Cultural changes manifest automatic upward transformations to change the truth value of cultural property to the state of its maximal truth value.

The cultural automaton has properties of automaticity that observe the characteristics of the cultural system. The cultural automaton may be a thing with a preset instruction to encode particular patterns of social and cultural thought. The patterns of social and cultural thought and behavior in cultural automatons are consistent with the lawlike regularity and principles of the cultural systems. Cultural systems guide patterns of social thought underlying behavior such as the social norms of societal organization.

Cultural systems provide guidelines for social rationale. Cultural systems determine heuristics for formal and intuitive social reasoning. Properties of automaticity in the cultural system observe characteristics that lead to social reasoning. Cultural systems show a distinction in the use of causal explanations that define the directionality of conceptual social relations. Cultural variation in causal social explanations defines the directionality of social relations in a dispositional or situational manner. Cultural systems that value dispositional social explanations place emphasis on social trait attributes. Cultural systems that rely on situation-based social explanations place the directionality of social relations on social trait attributes within the context of a social background.

Cultural variation in spontaneous trait inference is observed as a function of automatic social attribution. Cultural variation in spontaneous trait inference within the context of a social background illustrates the dependence of automatic social attribution on the context. Cultural variation in automatic social attribution shows the importance of cultural systems to affect the directionality of social relations. Cultural variation in social explanation is observed as cultural properties of automatic social thought.

Cultural changes in the superenvironment may affect components of social rationale. The input of cultural property in the superenvironment includes the use of automatic social attribution. Cultural changes as transformations of cultural property in the superenvironment serve to contribute to the highest levels of preparedness and protection. Cultural changes in the cultural property of the superenvironment manifest as automatic upward transformations to produce high levels of protection. Cultural changes in the cultural property as upward transformations of automatic social attribution guide epistemology of social knowledge. Cultural changes in the cultural property of automatic social attribution as upward transformations provide starting points of social inference. Cultural property that undergoes cultural changes of automatic social attribution reflects a focal point of social explanation.

The action of transformation of cultural property as automatic social attribution shows multifunctionality. Cultural systems may guide cultural property towards automatic transformations of social attribution that build the maintenance and transmission of cultural characteristics. Transformations of automatic upward social attribution enhance the truth value of cultural property in the cultural system. Cultural change in cultural property is efficacious to affect the directionality of social relations. Cultural property that initiates social inference transmits the characteristics that guide and build social reasoning. The cultural property in processes of social inference illustrates the value of the acquisition of social knowledge in the cultural system. Cultural systems rely on social rationale for the implementation of social norms that are foundational for societal organization.

Cultural change in patterns of social thought performs a function of reparation in cultural property. Automatic negative social attribution reflects the social perception of a necessary reparation of cultural property through automatic positive social attribution. Transformations of automatic upward social attribution show the capability to redress through the cultural characteristics of cultural property. Automatic positive social attribution illustrates a constraint satisfaction in the potential truth value of cultural property. Automatic positive social attribution in cultural property shows overall satisfaction or harmony in the pattern of activity that comprises cultural thought.

Cultural automatons may serve as a mechanism of cultural transmission of social attitudes. Cultural attitudes comprise a pattern of automatic social thought consisting of learned knowledge about people, places, and things in the environment. Cultural models that hold patterns of automatic social thought may transmit social attitudes to other minds through processes of social learning. The cultural transmission of social attitudes reflects the encoding and persistence of patterns of automatic social thought across individuals and groups.

The cultural transmission of social attitudes builds social relations and social identity. Positive social attitudes strengthen the social relations and social identity of ethnocultural groups (Berry, 2006). Automatic or implicit attitudes show the unconscious learning of social attributes in the cultural characteristics of cultural property. The automaticity of positive social attitudes reflects the acquisition of patterns in social thought that enhance the potential truth value of cultural property. Negative social attitudes suggest the importance of intergroup contact for social learning and the acquisition of social perception that builds and maintains social relations. The automaticity of negative social attitudes suggests the importance of conscious social thought to change patterns of social thought. The malleability of social attitudes constitutes a function of social regulation of intergroup relations. The cultural maintenance and transmission of positive social attitudes strengthen social relations and social identity.

In the superenvironment, cultural automatons are a source for the generation and transmission of knowledge and information for social learning. Cultural automatons contribute to the maintenance of cultural attitudes that strengthen the social identity and social relations of ethnic groups. Cultural automatons that generate novel social attitudes to enhance cultural property strengthen access to resources of protection and empowerment in the superenvironment. The generation of novel social attitudes may reflect the processes of social learning to produce specific changes in social or political action.

The cultural automaton can manifest in distinct physical forms. The cultural automaton in a digital environment consists of a digital representation, such as a still image icon or animation, that is associated with a set of things congruent with a pattern of social or cultural thought. Cultural automatons in the digital environment can manifest to transform input into output that is meaningful and recognizable as a social and emotional expression of the cultural system. The automatic production of functional output that has social significance and is trainable maintains and strengthens the cultural system. Cultural systems in patterns of social and cultural thought consist of component tasks. Given a particular input, the cultural automaton encodes the specific input to produce a preset functional output.

The cultural automaton in a physical implementation that is lifelike consists of the set of organisms and their environment that have properties of automaticity and can move or act by themselves without conscious or volitional control in the cultural level of the physical system in the physical world. The cultural automaton as a living superorganism can detect the potential of all of the automatic properties of the physical environment that comprise the component parts of its cultural system. The cultural automaton as a living superorganism can produce all of the possible automatic properties and component parts of the physical environment and cultural system. As a living superorganism, cultural automatons can coordinate the production of patterns of social and cultural thought that comprise component parts of the fulfillment of its potential. The cultural automaton can produce output in patterns of social thought that are producible based on its potential. The movement and action of the cultural automaton are based on the fulfillment of its potential.

The cultural automaton may also perform actions that maintain and reinforce cultural processes with instructions previously given to it by the computer. The cultural automaton includes preset instructions that are encoded or automatically performed with cultural actions that are previously given through computer instructions within the type of cultural system. For example, a cultural automaton may be a thing with a preset instruction with programmed streams of automatic cultural thought. For cultural dimensions, programming a cultural automaton to produce automatic cultural thought entails physical state transitions. The computer that performs cultural computation may give instructions to the cultural automaton for the automatic production of cultural actions.

Cultural Automatism in Mental Computation

Cultural automatism of biological organisms refers to the information-processing mechanisms that perform automatic processing of features of the organized system at the cultural level. Cultural automatism consists of the mental and physical properties of cultural processes that occur in an automatic manner from mental computation. Mental computation of cultural processes is the set of neural information-processing mechanisms that perform mental functions to generate cultural patterns of automatic thought. Mental computation of cultural processes consists of cultural neural networks that perform basic computations based on patterns of network activation.

Cultural patterns of automatic thought refer to the stream of experience that is produced in the automatic processing of sense data from environmental input. Cultural patterns of automatic thought consist of basic mechanisms that produce simple responses from sense data based on autoassociation. Functional specialization of mental modules is comprised of information-processing mechanisms for the mental performance of automatic thought.

The functional specialization of information-processing mechanisms tunes to the cultural sense data from the environment. Cultural sense data is part of the specialized input–output relation in the automatic processing of streams of thought. Cultural sense data as an initial state of the specialized input–output relation consists of the configuration of features as an arrangement of parts in the cultural level of the organized system. The cultural sense data acts as the cultural input to an information-processing mechanism that automatically produces accurate cultural output.

Cultural sense data consists of the features as parts of the initial states of complete patterns of activation in the network. Cultural sense data is a range of initial states to a set of complete patterns and final states of cultural neural networks. The amplification of features as the activation of initial states of complete patterns leads to the same particular final state as the accurate cultural output. The functional specialization of the input–output relations is comprised of the set of the complete patterns and a range of initial states as the sense data of the cultural level in the organized system.

Cultural neural networks consist of the biological basis of cultural processes. Cultural neural networks are comprised of interconnected networks of neural activation that perform specific mental functions. Cultural neural networks demonstrate patterns of neural activation based on the functional connectivity of brain regions. The functional connectivity of brain regions contributes to the causal flow of information as an activation pattern across the cultural neural network. The activation patterns of cultural neural networks entail complete activation patterns that lead to final states from a range of initial states.

Computational principles of the structural and functional components of the nervous system contribute to the automatic and controlled processing of information at the cultural level. Automatic processing in information-processing mechanisms is considered lower-level, while controlled processing of information is thought of as higher-level. Automatic processing of information refers to learning algorithms in mental computation, such as classical conditioning, that pair a simple stimulus with a basic response. Automatic information-processing mechanisms consist of simple feedforward nets that demonstrate unidirectional feedforward processing.

Controlled processing of information consists of functional tasks, such as executive function or inhibitory control, that control the processing of information through representation or interpretation within hidden layers of the neural network. The regulation of automatic information-processing mechanisms demonstrates the control of automatic responses through inhibition or reinterpretation. The regulation of automatic thought entails bidirectional connectivity in neural networks with hidden layers for the control or inhibition of automatic information processing to a response.

Cultural Automatism in Machine Computation

Cultural automatons are the set of things in the world that are autonomous and self-organizing at the cultural level of the physical system in the physical world. Cultural automatons demonstrate the cultural automaticity of the computing machine. The computing machine describes the automated production of deterministic output based on input–output mappings. The computing machine entails a range of types of computation from the mental computation of the biological organism to the machine computation of the cultural computer.

Cultural automatism refers to how computers and devices automatically produce information content at the cultural level of the physical system. Cultural automatons that demonstrate autonomous behavior have the capability to move or act by themselves. Cultural automatons are an example of the performance of basic functional tasks as automated machine computation. The automated production of cultural informational content, including cultural programs, describes cultural automation in the physical system. Cultural automatons perform the automated actions of the cultural level in the physical system.

Cultural automatons that show the functional equivalence of machine computation and mental computation contribute to the parts and particulars of the artificial living system. As artificial life, cultural automatons display agentic characteristics that are a part of the automated production of machine computation (e.g., robotic arm, synthetic brain). The design of cultural automatons includes the simulation and construction of the reality of real-world experience in synthetic devices and neurotechnology that are close to real anatomy and physiology. Nevertheless, the incommensurability of mental and machine computation in other mental characteristics, including intentionality and experience, entails the distinction of machine computation in artificial life from the mental computation in mental life.

The automated production of machine computation in artificial life contributes to the restoration and reproduction of natural life. The artificial intelligence approaches to the production of artificial life range from the simulation of mental functions through neurotechnology to the construction of synthetic devices that produce basic functions. The construction of synthetic devices that produce basic functions builds on the application of integrative neuroscience to neurotechnology.

Conclusion

Automatism as a mental and physical property in the physical system of possible worlds presents novel and interesting philosophical considerations. Automatism contributes to the computational processes in minds and machines. Computational models describe the functional architecture of systems that demonstrate the automated production of information processing. Automatic processing of thought complements the mechanisms of controlled processing or higher-level mental processing. Understanding the levels of processing of the computational cultural brain contributes to broader philosophical understanding of the nature of the mind.

Reference

Berry, J.W. (2006). Contexts of acculturation. In Sam, D.L. & Berry, J.W. (Eds.). *The Cambridge handbook of acculturation psychology.* New York: Cambridge University Press.

Further reading

Godfrey-Smith, P. (2003). *Theory and reality: An introduction to the philosophy of science.* Chicago, IL: University of Chicago Press.

3

INTERFACE THEORY

Introduction

The philosophical notion of the mind as a computing machine that produces mental thought presents a range of questions regarding the nature of the mind and its functional and causal role in possible worlds. The postulation of the mind as a source of mentality and thought from mental computation suggests that the mind and its features are a putative model for the design and construction of mental content. The mind, consisting of mental properties that perform specific functional and causal roles, addresses classic questions of the nature of the mind. The mind as a model for the computing machine implies that the machine consists of machine properties that contribute to the functional role of mental content in the world.

The design and construction of mental content from machine computation suggest numerous considerations regarding the functionality of mental content for machine computation. Mental content from machine computation implies that machine content is a type of machine property. Machine content is a source of machine capital or the valuation and worth of the machine property. The valuation and worth of machine capital are based on a standard of criteria consistent with the purpose and functionality of machine property.

The design and construction of mental content from machine computation contribute to one of the earliest functional roles of machine computation for social communication. One of the early functional roles of machine computation is to produce machine property consistent with that of the mental property produced with mental computation. Machine property that facilitates social communication suggests a standard of criteria for machine capital that is consistent with the multiple physical realizations of mental content.

The mental capacity for social recognition implies the importance of social representations in mental computation. The design and construction of mental content from machine computation require the capabilities for social recognition of a range of social representations. Machine computation for social communication entails the design and construction of a computer system for the automated production of mental content for social communication.

The construction of a computer system for the automated production of mental content for social communication requires the design of a computer interface. The computer interface acts as a point of contact between minds and machines. The design of the computer interface built from the features of mental computation implies the existence of a point of contact between mental and machine computation or between the causal systems of the mind and machine. The computer interface acts as a point of contact for the interconnection of mental and machine computation.

Interface theory introduces the notion of the interface in philosophy of mind as point of contact between mental and machine computation as a deterministic boundary across two parts of matter or two systems of possible worlds. The philosophical notion of interface theory entails the intentional design of a point of contact between minds and machines. The design of a point of contact between minds and machines is consistent with a notion that there exists a commonality in the functionality of minds and machines. The commonality of functionality in minds and machines suggests a functional role for complementary interaction.

Interface theory postulates that the nature of the mind is bounded and deterministic at the point of contact with the computing machine. The interface between minds and machines is deterministic and existent within spatiotemporal dimensions. The point of interaction of mental computation and machine computation acts as a boundary to possible worlds. Interface theory ensures that there is in existence a boundary between mental and machine computation. Interface theory as a philosophical position posits that computation occurs across distinct spatiotemporal dimensions within the mental and physical world that are interconnected across a boundary as a point of contact.

The design of the computer interface is the development of system capabilities that brings into existence a point of contact for interaction across two systems in possible worlds. The development of system capabilities of a computer allows the interaction across mental and machine computation or the interaction of mental content across different environments. The interaction of minds and machines suggests that there exists a standard of criteria that defines mental content across multiple physical realizations. The interaction of two systems in possible worlds further implies that that there is an actuality to the relation of the two systems with directionality or causation. The interface as an existence of the interaction of minds and machines brings the performance of real-world experience into novel possible worlds.

In philosophy of mind, interface theory is the stance that the existence of a point of contact across two parts of matter or two systems in interaction matters for mental content. Interface theory implies the scope of mental content relies on the existence of a boundary across two parts of matter of possible worlds. The philosophical notion from interface theory is that the interaction of two parts of matter forms a boundary as a point of contact. From the interaction of two parts of matter is a point of settlement, a deterministic limit that acts as a boundary of matter. Interface theory also entails the existence of a point of contact of inter-action across two systems in possible worlds. Interface theory is a philosophical account of the implication of the point of interaction across causal systems. The causal systems in interaction with one another may consist of a particular level of an organized system.

In philosophy of science, interface theory implies that the commonality of functional role of mental and machine computation entails the existence of basic patterns and principles in computation. The existence of the interaction of minds and machines suggests an actuality to the fundamental principles of computation that subsumes mental events. Interface theory as a postulation of fundamental principles of computation is consistent with unification theory that assumes that basic principles connect sets of facts in the world.

Interface theory posits that computation as a science contributes to the philosophical inquiry of explanation. The formulation of a point of contact as interaction of minds and machines suggests causal relations among mental and machine computation. The interaction of parts and particulars of mental property as machine property and machine property as mental property contributes to the mind as a causal system.

Minds and Machines

In philosophy of mind, the mind defines a standard of thought in mentality and intelligence. The properties of the mind consist of sets of mental states as mental events. Mental properties as features of the mind define a model of computation from the biological machine. The mind as a biological machine suggests that mental features perform specific functional roles that contribute to adaptation.

The mind as a computing machine connotes the notion of mental thought as a set of mental properties. The mental properties of thought consist of the set of mental events that occur as a pattern or regularity in the world. Mental properties consist of the mental states of mentality and thought. Mental state inferences are properties of the mind that are generated in the interaction with environmental input. Mental state inferences arise from the input of sense data into information-processing mechanisms that perform transformations into representations as interpretations that produce an appropriate response. Mental state inferences comprise the output from environmental input into the bio-logical computing machine.

The philosophical notion of the mind as a computing machine places emphasis on computation as a fundamental basis of the mind. Mental computation describes the computational basis of the mind and its functional properties. Mental computation is the performance of mental functions for problem solving. The generation of mental state inferences from a set of component tasks comprises the computational level of analysis. The computational level of analysis entails the decomposition of specific problems into component tasks for the production of a particular solution. The generation of mental thought for solving problems in the environment constitutes adaptation.

The computer as a simulation of the mind is machine computation for the production of mental content. Machine computation includes the design and construction of computer programs for the production of mental content. The programmable functions of the computer consist of a set of functions that define formal procedures or sets of rules for information production. Machine computation is the production of mental content from a set of programmable functions. Mental content from machine computation demonstrates the biological plausibility of thought from basic mechanisms.

The mental content from machine computation constitutes a form of machine property. Machine property as mental content demonstrates the functionality of computer programs for social communication. The mental content of machine property consists of the automated production of mental state inferences. The machine computation of mental state inferences implies production of mental content from rule-based structures.

Machine computation illustrates the multiple physical realizations of the mind. Machine computation implies that the production of mental content has multiple physical realizers. The multiple physical realization of the mind is the mapping of the set relation of mental states to multiple physical states. The mapping of a mental state to a physical state entails that the identity relation defines the functional operation of mental and physical properties in state space. The mapping of a mental state to multiple physical states suggests that mental and physical properties in state space are defined by multiple functional operations.

The properties of mental content from mental computation and machine computation show the functional equivalence of computation. The interaction of minds and machines implies the existence of a functional equivalence in the production of mental content. The existence of minds in interaction with machines is the performance of mental computation for machine computation. Mental content from mental computation is the mental property that can supervene on machine property. Mental content from machine computation is the machine property that can perform the functional role of mental property.

Mental Computation and Machine Computation

The causal interactions of minds and machines are multifold. Mental computation as the production of mental content for machine computation implies a causal relation between mental and machine computation. The interaction of minds and machines as the supervenience of mental property on machine property implies the production of machine property from mental property. Mental property plays a causal role in the automated production of machine property and the transformation of machine property into mental property. The causal interactions from mental and machine computation suggest the bidirectionality of information flow across mental and machine property. The bidirectionality of information flow across mental and machine property is an example of the interconnection of mental and machine computation.

The complementarity of mental and machine property implies a functional role of mental causation in mental content. Mental property as a source of supervenience plays a causal role in the production of machine property of possible worlds (e.g., the mind in the world). Machine property as a source of environmental input plays a causal role in the representation and transformation of sense data into mental property (e.g., sensory representation in the mind).

The interaction of minds and machines suggests a range of issues that arise from mutual influence. The commonality in the functionality of mental and machine computation implies a multitude of sources for mental content that fit a standard of criteria in truth correspondence with the world. The interconnection of mental and machine computation suggests the possibility for commutation of mental and machine property as part of the discovery of truth theories in the world. Mental property is a discovery process of the truth correspondence of patterns and regularities in the natural world. Machine property is a design and construction of mental content in correspondence with the truth theories in the structure in the world.

Mental and machine property contribute to computational discovery as a resource that informs explanation. The interaction of minds and machines is a form of social communication that brings the real-world experience of the physical world into the immersive reality of digital and virtual worlds. Mental property as machine property is the design and construction of tools for the augmentation of reality that enhances the societal perception or the actuality of the physical or immersive environment. Mental property as machine property contributes to the automated control and production of mental content from the physical environment into digital or immersive environments. The intentionality to design computer systems that perform automated production is an example of mental property as a source of machine property. The interaction of minds and machines allows machine property as part of the sense data that informs theory building to a satisfactory and complete explanation.

Machine and mental property demonstrate trade-offs in the automation and control of information production. Machine property as mental property is the performance of programs that simulate the augmentation of reality and produce mental content. The augmentation of reality is an improvement when the production of mental content from machine property performs a causal-functional role that is necessary and sufficient. The automated machine production of programs in devices and computers is an example of a simulation of the augmentation of reality from machine property. Machine property as mental property is a tool for the automation and control of mental content from digital or immersive environments into the physical environment. The use of computational discovery to inform the truth correspondence of mental content with the structure of the world is a tool for scientific and technological advancement.

The incommensurability of mental and machine computation places limitations on the commutation of mental and machine property. Mental computation comprises knowledge generation for explanatory inference; machine computation implies knowledge extraction for prediction. Mental computation entails knowledge generation of mental content from the biological computing machine of the individual; machine computation entails the information production of mental content from the computing machine of the group. These limitations in the commutation of mental and machine property represent multiple constraints on the functional equivalence of computation.

Culture and Mental Computation

Cultural processes affect the generation and maintenance of mental content for cultural and psychological adaptation. Cultural processes guide the cultural transmission of social information across individuals and groups. The cultural transmission of thought demonstrates patterns and regularities that are consistent with cultural selection. Cultural patterns of thought guide the social norms and societal organization of the group.

Mental property is part and particular of the properties that create and maintain the cultural level of living systems. Cultural mental content consists of the set of component tasks that perform problem solving for the cultural group. Cultural mental content consists of the mental state inferences that contribute to the social norms and societal organization of cultural groups. Cultural mental content comprises the mental thought that is produced from the specialized functionalization of the mind in the cultural group. Cultural processes describe the generation of cultural mental content based on culture-based sets of rules. The generation of mental thought that strengthens the social norms and societal organization of the cultural group constitutes cultural adaptation.

Cultural mental property consists of the mental features that define a cultural model of computation. The cultural mind as a biological machine entails features of the mind that perform specialized functional roles. The functional specialization of mental thought is based on the differential tuning of information-processing mechanisms to the sense data of the cultural group. Cultural patterns of thought strengthen features of the mind that contribute to psychological and cultural adaptation.

Cultural models of computation refer to the sets of component tasks that perform problem solving within the cultural context. The cultural context provides culture-based sets of rules for the transformation of culture sense data into informational representation that produces cultural behavior. Culture-based sets of rules define algorithms for the production of culture-specific behavior. Culture-based learning algorithms entail the production of culture-specific output across multiple physical realizations.

Cultural models of mental computation define the task components for problem solving that contribute to cultural and psychological adaptation. Cultural models of mental computation include mental constructs of emotional, cognitive, and social domains and their specific functional tasks. Cultural models guide the transformation of sense data into internal representations across distinct domains. The physical implementation of mental constructs consists of neural information-processing mechanisms for cultural neurocomputation.

The real-world experience of the mind constitutes a state of consciousness that is private, bounded, and unique. The capacity of the mind to generate cultural experience in the stream of consciousness demonstrates the functional adaptation of the biological machine for culture. Cultural experience as conscious mental states reflects the causal-functional relations of the mind and brain at the cultural level. Culture as conscious experience comprises the parts and particulars of cultural mental content in the physical world.

Culture and Machine Computation

Machine computation contributes to information production at the cultural system level. The design and construction of machines for the production of informational content include the programming of functions of the system that contribute to cultural production. Machine computation at the cultural system level demonstrates automated control of the production of informational content. The design and construction of devices and computers that perform machine computation are examples of cultural advancement through scientific and technological progress.

The production of informational content of the cultural system in machine computation includes programming functions that perform at a level of functional equivalence as mental computation. Machine computation requires the programming of functions that respond to cultural mental content. The programming of the computing machine for cultural performance consists of the design of features that enable the mind to control the automated production of cultural content. The design of user-based cultural scripts and cultural scenarios allows the user to guide the performance of culture-based functional tasks.

The programmability of cultural functions implies a role of intentionality and control in the production of cultural content. The design of user interfaces that build the societal influence of particular cultural content contributes to cultural dynamics. Cultural dynamics describe the frequency, prestige, or popularity of mental content among cultural groups. The use of cultural content as machine property demonstrates a causal-functional role of the cultural content from machine computation in the cultural dynamics of social groups.

Machine computation demonstrates that the machine performance of cultural tasks is produced through programmable functions. Machine computation provides a functional architecture amenable to the programming of functions for cultural performance. Culture-based rule sets define algorithmic levels of analysis in machine computation. The sets of functional tasks for cultural performance describe the computational level of analysis.

The automated production of cultural content from machine computation is a source of sense data in mental computation. Cultural machine content acts as an environmental input in mental computation. Cultural machine content is produced to complement or to supplant the functional roles of mental computation. Cultural machine content guides societal perception through the shaping of cultural patterns of thought.

The machine computation consists of the production of informational content as the machine property of the cultural system. Machine computation at the cultural level implies that there exists a standard of criteria for the production of cultural content. Cultural machine content demonstrates the capability of machines for knowledge extraction from datasets and prediction of cultural patterns and regularities in the world. The use of technology as a tool for the shaping of cultural patterns of thought demonstrates the societal influence of the digital and immersive environments on the cultural level of organized systems.

Effective social communication between humans and computers requires the design of interfaces that understand the cultural context. Cultural competence is the capacity to share the social communication cues that are consistent with display rules and social norms of the cultural group. Social communication that shows cultural competence benefits the societal coordination of resources and the attainment of goals for individuals and groups. Social characters in digital and immersive environments that show cultural competence in the performance of social scenarios build the social influence of the cultural context. The development of technology as a tool of prevention and intervention facilitates the training of cultural competence across possible worlds.

The interaction of humans and computers for cultural performance describes the causal-functional relation of minds and machines at the cultural level. Computer programs provide a tool for cultural development by enabling the simulation of cultural participation in possible worlds. The real-world experience of culture as states of consciousness broadens the scope of cultural participation and the actuality of culture in the physical environment. The control of the automated production of cultural content across possible worlds demonstrates the impact of machine computation on cultural advancement.

Cultural Computation

Cultural computation consists of the functions of independent and interactional properties of computational components at the cultural system level. The cultural computation of minds and machines in functional operation as independent properties describes parts of the cultural level of an organized system. The cultural computation of minds and machines in interaction describes causal-functional relations in computation at the cultural level. The causal interactions of mental and machine computation demonstrate the interconnection of minds and machines.

The directionality of causal relations of mental and machine computation demonstrates trade-offs at the cultural level. The interconnection of real-world experience of mental computation into the digital or immersive environment suggests a point of contact between the causal system of the mind and its interaction with the machine. The input of mental content from mind to machine implies a causal influence of mental computation on machine computation. The mind as a causal actor assumes a level of responsibility and acts as a guide for the arrangement of events in the digital or virtual world. The content of the mind as a causal factor is the source of input for information production from the computing machine. The mind is responsible for the mental content of experience and reality in the digital or virtual world.

The mind as a guide for the arrangement of events implies the existence of features of mental states in the digital or virtual world. The interaction of the mind in the machine entails a cultural milieu of social interaction. The cultural mental content in the machine consists of a system of features that comprise the cultural elements of digital and immersive environments. The mind as a guide for the arrangement of cultural events implies the necessity for knowledge of the features of the digital and immersive environment. The mind assumes there exists a level of responsibility that controls the machine performance of social representation in digital and immersive environments and its spatiotemporal properties. The mind assumes a level of responsibility in the control and automation of the machine performance from social representations, such as agents and avatars.

The mind builds the features of the digital and immersive environment through social interaction. The mind as a causal actor creates varieties of social representations in the digital or immersive environment, such as creating a virtual network or a virtual identity. The mind as a causal agent designs the features of the digital or immersive environment through the specification of parameters that control events and event objects. The specification of digital events and digital event objects as the basic units of functions defines the output of programs.

The transition of mental content from causal actor to causal agent entails a causal influence of mental computation on machine computation. The causal influence of mental computation on machine computation is a change in the units of matter for which the mind assumes a level of responsibility. The mind as a causal agent guides the arrangement of events in the digital or virtual world through devices and computers that perform an augmentation of experience and reality. The mind as a causal agent controls the automated production of events and event objects in the digital and immersive environment. The design and construction of tools for the automated production of machine performance act as an interface between minds and machines. The interface as part of the programs of devices and computers is the point of contact from which the mind controls the automation and production in the environment.

The concept of the interface of minds and machines is important to philosophy of science. The design and construction of computational tools provide the language and cognition from which to understand the structure of the digital and virtual world as possible worlds. The interface is a source of the set of concepts from which the mind as a perceiver observes the digital and virtual world. The mind is the complete source of explanatory inference and reason in interaction with the machine. The mind is a source of prediction from mental performance; the machine interface is a source of prediction from machine production. The machine interface is a source of the production of machine performance as a set of operations ranging from the probabilistic automaton to deterministic automaton. The programmable functions of machine performance demonstrate the importance of intentionality in the state transitions of machine production. The mind and machine in interaction act as multiple representations that perform in the digital and immersive environment. The mind as a causal system is the source of the complete and satisfactory explanation of events across possible worlds.

Conclusion

The concept of an interface between minds and machines implies an interconnection through computation across possible worlds. The use of machine computation for the production of mental content places importance on the philosophical notion that the mind is a cultural model of computation. The mind as an ideal model of computation is a fundamental source of real-world experience in organized systems. The computational modeling of the mind provides a means for the automated production of mental content. The automated production of mental content contributes to the conservation of the mind and its functions as a causal system.

The interaction of minds and machines contributes to the understanding of explanation in philosophy of science. Traditional notions of explanation attribute the mind as a source of explanatory inference. The use of tools such as computers and devices for the production of mental content introduces a secondary source of prediction. The use of machine computation for prediction demonstrates the pertinence of computational discovery as evidence in theory confirmation. The computational discovery from machine computation contributes to the mind as a complete source of explanation and scientific reason.

4

MACHINE FUNCTIONALISM

Introduction

The conceptual foundations of the cultural mind consist of an understanding of the mind as a causal system. The conceptualization of the mind as a causal power implies the existence of mental events as a network of causal events. The mind as a causal system illustrates the complexity of mental events as physical events in the physical world. Mental events as a causal power of the mind demonstrate a real-world consequence that is consistent with the functioning of physical systems. The complexity of the mind as a causal system is a consideration in parallel to the complexity of the functions of physical systems.

The historical significance of machine functionalism is a philosophical inquiry into the nature of the mind in a physical system comprised of specific functions. Early philosophical notions of machine functionalism conceptualize the mind as consisting of mental property. The notion of mental property posits a realization of the mind as physical property in a physical system. Mental events are physical events characterized by physical properties with functional causal roles.

Machine functionalism highlights the notion that the mind as a machine is capable of functional performance. The notion of the mind as a machine connotes the functional role of mental property as a causal mechanism. The mind as a machine assumes that mental events perform a functional task consistent with transformations of input and output. The mind as a machine performs tasks as with its physical instantiation.

The early conceptualizations of machine functionalism depict the mind as consisting of mental properties. For the mind, mental property as physical property defines mental states as physical states consisting of the spatial and temporal properties of a physical system. Mental states as physical states with spatial and temporal properties show the capability of functional performance as a physical system. The mind as a physical property with performance as a physical system can assume distinct functional causal roles.

Mental property can occur in multiple physical realizations. The mind can assume multiple physical realizers of distinct physical properties. Multiple physical realizations of the mind imply that the primary functional role of mental property is consistent across its physical instantiation. The physical realization of the mind is characterized by its functional role. The physical properties of the mind exist to perform specific functions. Mental property as physical property fulfills a functional causal role.

The importance of machine functionalism for the computational mind is the emphasis on computation as the causal power of the mind. The causal power of mind arises from its computational performance. The causal power of the mind for mental representation as an organized set of causal mental events illustrates the complexity of the mind as a causal system. The causal power of the mind for transformation of mental property into causal mental events further demonstrates the functional performance of the mind.

The importance of the notion of machine functionalism for understanding culture is multifold. Machine functionalism introduces the concept of culture as a causal system of the cultural mind. The generation and construction of culture are the production of cultural property from a cultural network of causal mental events. The cultural mind as a causal mechanism illustrates the causal power of cultural mental events. The network of cultural mental events consists of physical events with spatial and temporal properties. Cultural change reflects a rate of change in the spatial and temporal properties of cultural mental events.

The cultural mind as a cultural computing machine underscores the functional role of culture in a physical system. The cultural mind as a causal power performs distinct sets of cultural computations as a network of causal events. Cultural mental events are instantiated as physical events for specific functional use. The cultural mind as a causal system can assume multiple physical realizations, from the cultural brain to the cultural computer. Across physical realizers, the cultural mind consists of functional causal roles for the performance of cultural tasks.

Machine Functionalism

Machine functionalism as a stance in philosophy of mind is a description of the mind as a computing machine of specific functions. The mind as a computing machine performs specific functional roles based on task components. Machine functionalism places the emphasis of the mind on the performance of functional roles. In machine functionalism, mental events are causally related to physical events with specific functional performance. Mental property is the set of mental events that are performed to fulfill specific functional roles.

The multiple realization of mental property in physical realizers underscores the importance of the functional specialization of the mind. Because mental properties have more than one physical instantiation or physical identity, machine functionalism considers the function or purpose of the mental property as a primary characteristic of the mind and its mental property. The common consideration across multiple realizers is the functional purpose of the mental property. Functional properties of mental content can have multiple physical realizers or multiple physical instantiations.

The multiple realization of mental property in a physical system suggests the further possibility that physical instantiations of mental property can serve multiple functional roles. That is, mental property consists of the set of mental events which are linked to the set of multiple physical events. For each mental event, there exists a set of physical events that are realizations of the same mental event. For each physical event, there exists a set of mental events that have a common, specific physical realizer. Thus, the principle of one-to-many mapping is characteristic of the mental and physical property in machine functionalism.

Machine functionalism is the mind in performance of functional operations as a computing machine. The function of a given mental property is to perform a specific causal role. Causal functionalism is the postulation that mental property performs a specific causal function. Machine functionalism is a kind of mental realism, that is mental property that performs a specific causal role in the real world. Mental property as a network of mental events performs for a specific causal function or set of causal functions.

Machine functionalism requires consideration of the physical device as a realizer of the mental content of the machine state. The notion of the mind as a computing machine requires a demonstration of the capability of the construction of a physical device for mental task performance. The Turing machine is a simple demonstration of the construction of a physical device for the performance of mental output (Turing, 1950). The Turing machine is an example of the mind as a computing machine in a physical device.

The Turing machine consists of a physical device with multiple components. The Turing machine is a physical device with four components, consisting of a tape, a scanner-printer, a finite set of configurations of internal states and a finite set of alphabet symbols. Given a particular input, the Turing machine produces machine output that is consistent with mental output. The machine output of the Turing machine, for instance, consists of a string of symbols. The input-to-output performance of the Turing machine suggests that a physical device can demonstrate the performance of mental computation. The machine shows the capability of symbolic operation, for instance, to print or to add a symbol or delete with another. The machine can produce a string of symbols as an output based on preset instructions or rules specified in a machine table.

Consistent with the notion of the mind, the Turing machine relies on a configuration of internal states for the production of machine output. The Turing machine as a physical device for input–output performance is based on physical states and its transitions as internal states of the machine. The internal state of the machine is the physical state of the machine; the internal state consists of mental content that can be edited based on specific rules or instructions in the machine table. The input-to-output performance of the Turing machine is based on physical state transitions that consist of changes to the internal states of the machine. The Turing machine furthers the notion of computation as a set of internal events in a physical device.

The Turing machine as computational device demonstrates the importance of rules for the performance of mental output. The identical performance of machine output can be achieved through distinct machine tables. That is, different sets of rules or instructions can produce the same string of symbols or machine output. The use of different rules as machine tables is sufficient for the recognition of distinct machine identity. Machine input–output equivalence does not necessarily imply the same mental computation. Thus, machine input–output equivalence does not necessarily imply the same Turing machine.

Machine functionalism demonstrates a functional role of machine computation for deterministic production. Deterministic machine functionalism shows that a computing machine produces a specific predetermined output given a particular input and machine table. The computation of the machine is deterministic for the production of a given output. Machine computation produces programmable rule-based output that is deterministic.

The determinism of machine functionalism suggests the relevance of machine computationalism as a universal machine. The universal machine can be programmed to perform the computation of any machine table. The computing machine is capable of computation based on any set of programmable rules. The machine is capable of demonstrating machine output that is general-purpose, rather than functionally specific.

The notion of culture as a universal machine reflects the programmable functional operations of the computing machine and implies the automated production of the property of minds and machines at the cultural level. Cultural mental content of the computing machine in automated production is defined from rules of mental computation for machine production. Culture as shared meaning systems consists of rule-based computation and the performance of cultural rules of transformation for cultural production. The automated production of cultural mental content based on programmable cultural rules demonstrates a deterministic source of cultural production. Culture as a universal machine contributes to a standard of criteria in the hierarchy of universals (Norenzayan & Heine, 2005) or levels of universalism in machine functionalism.

Probabilistic machine functionalism shows the capability of machine computation for probabilistic production of final states. As a probabilistic automaton, the computing machine can produce specific output through instructions based on probabilities. The production of specific machine output can be preset to a given probability. Probabilistic automatons suggest that machine output can produce probabilistic output based on the input of probabilities as physical state transitions. Probabilistic automatons show that probabilistic output is a type of machine output that is not necessarily inherent given a particular input, but instead rule-based and programmable through computation, such as in a machine table. The input–output performance of probabilistic automatons suggests distinct machine identity based on the probability of output, rather than the specification of rules per se.

Culture as a universal machine consists of the programmable functions and machine table for the automated production of mental content as a probabilistic automaton. The notion of a cultural probabilistic automaton refers to the automated production of cultural mental content that is probabilistic or stochastic as an adherence to the programmability of functional operations. The automated production of cultural mental content of a cultural probabilistic automaton relies on functional performance as internal states that are fixed and stable. The functional operation of the computing machine as a cultural probabilistic automaton can perform a simulation as a computing machine that produces probabilistic output as deterministic. Cultural production as a probabilistic automaton implies the functional operations of the computing machine for probabilistic output or the simulation of the performance of the computing machine as probabilistic.

Machine functionalism demonstrates that the mind as a computing machine performs specific functional roles for the production of symbolic culture. In the example of simple machines, the computing machine performs the functional role for the production of symbolic output. The performance of symbolic output in computing machines illustrates a functional role for computation in culture. The machine that performs deterministic or probabilistic symbolic production as machine output demonstrates how mental computation is a form of cultural production.

Mental computation is fundamental to culture. Cultural production from simple machines to computational devices highlights the role of mental computation for cultural performance. Cultural symbols are the production of mental computation based on functional role. The use of distinct rules for cultural production is sufficient for the recognition of distinct cultural identity and highlights the importance of cultural identity for cultural performance.

Culture as a universal machine demonstrates how mental computation is a causal power of cultural variation. The functional operations of the computing machine and the universal programmability of its functions based on different sets of cultural rules define the standard of criteria for levels of universalism in machine functionalism. For functional universals, mental computation varies across cultures based on the strength of the mental performance for a specific functional task. Mental computation as states of mental content that differ in strength across cultures produces cultural variation.

Machine functionalism in the cultural mind demonstrates the complexity of mental causation. The complexity in the cultural computing machine is depicted in the specification of general rules for the operation of the cultural machine. Cultural task performance demonstrates machine functionalism at the level of computation. Cultural variation in the performance of specific mental tasks illustrates the distinct functional use of mental property across cultural contexts. Cultural variation in mental content highlights the importance of functional role in mental performance.

Machine functionalism in the cultural mind performs functional operations as a form of cultural computation. The cultural computing machine consists of cultural mental property comprised of a set of cultural mental events with multiple physical realizers. Cultural mental property is the set of cultural mental events that have a causal role in cultural functions. The functional performance of cultural tasks consists of the mental content for cultural computation.

The cultural mind as a physical realizer in machine functionalism is as a cultural computing machine. The cultural mind as a cultural machine performs operations based on a specific table of instructions; the cultural machine has internal states with mental content through operations defined in the machine table. The cultural machine is a producer of the states of symbolic culture as cultural output. The performance of the cultural machine for the production of symbolic culture is a demonstration of the cultural mind as a causal power.

Mind, Brain, and Machines

From simple to complex machines, multiple physical realizers can perform mental computation. Simple machines as a producer of machine output demonstrate the capability of simple devices to produce mental output. The Turing machine places the causal power of mental events in the computation of simple machines. Complex machines, such as biological machines, demonstrate the biological plausibility of machine computation of the mind in the brain. The biological machine situates the causal power of mental events in the neural computation of the brain.

Mental property is a mental event with causal power in the physical world. Mental property plays a causal role in the production of mental computation. Mental property performs specific functions based on a given task. The functional property of mental states is to ensure either a deterministic or probabilistic causal output given a specific input.

Mental property is a type of mental event with causal power. To have a specific mental event is to cause a particular outcome. The outcomes of mental events are occurrences in the physical world. For instance, behavioral outcomes or behavioral expressions of the organism are a causal consequence of specific mental events. The occurrence of behavioral outcomes assumes mental causation.

In machine functionalism, mental events are in causal relation to physical events of the brain that perform specific functional roles. The realization of mental property as a physical instantiation in the brain illustrates the biological plausibility of mental computation. The functional properties of mental events have a physical instantiation as a biological machine. The physical states of the brain in the biological organism demonstrate the functional purpose of mental property.

The causal relation of the mind and brain is an instantiation of the lawlike regularity of natural phenomena. Mental states supervene on the physical states of the brain. Mental property in the physical realizer of the brain consists of the physical properties of brain states. The psychoneural identity of mental and physical events demonstrates the robust pattern of mental computation in the brain. The physical states of the brain are necessary and sufficient for the occurrence of mental states. The mind is a causal power of the physical state of the brain.

The functional property of the mind as a biological machine is the performance of functional operations as brain states. The function of mental property to perform specific causal roles is as physical states of the brain. The function or purpose of mental property is a primary characteristic of the physical states of the brain. Mental states of the mind are physical realizers in states of the brain.

The functional operation of the mind as a biological machine is a demonstration of mental realism. The mind performs specific causal roles in the real world as a biological organism. The brain of a biological organism shows the functional purpose of mental capacity for real-world advantages. The brain as a biological machine illustrates the robust patterns of biophysical mechanisms and their causal effects. The biophysical mechanisms of the brain implement neural operations that perform specific computations. The brain of a biological organism consists of the emergent properties of a physical system without a lower level of explanation.

The mental property of the mind consists of the physical property of the brain. For each mental event, there exists a physical event or set of physical events that are a realization of that mental event. For each physical event, there exists a mental event or set of mental events that share a functionally specified physical realizer.

Patterns of neural network activation comprise mechanisms and causal interactions among their parts. The principle of one-to-one mapping is characteristic of the functional mapping of a mental state to a brain state based on a specialized function. The principle of one-to-many mapping is characteristic of the functional mapping of a mental state to a network of brain states based on a specific function.

The functional architecture of the nervous system is distinguished across levels. Levels of organization refer to the functional anatomy of the nervous system, including the hierarchy of structural components that comprise neuroanatomy. Levels of processing refer to the spatial location of physiological mechanisms for information processing in the nervous system. Levels of analysis refer to neurocomputation as the functional performance of the nervous system.

The organization of the nervous system describes the functional architecture of the brain across spatial scales. The functional architecture of the brain demonstrates the specialization of function of sensory systems in the nervous system. For instance, the sense data from the environment consists of sensation transformed into the representational content of percepts encoded and stored in primary sensory regions of the brain. Across functional neuroanatomy, regions of cortical and subcortical lobes consist of specialized processing pathways to perform specific mental functions.

The large-scale specialization of brain areas is consistent with basic principles of computation. Cortical information processing consists of a hierarchical structure of specialized pathways. Information processing is distributed across a range of distinct specialized processing pathways. Layers of processing in the hierarchical structure require the performance of specific transformations. The information processing of cortical layers is dedicated and content-specific to patterns of activation. The information processing of activation patterns is deterministic to the specific properties of stimulus, rather than general-purpose.

Across levels of processing, the organization of the nervous system demonstrates functional specialization. Levels of processing refer to the relation between neuroanatomy and information processing. Distance from sensory input to functional neuroanatomy reflects higher levels of information processing. Parallel streams of processing characterize the projection of sensory information across cortical regions. Bidirectional projection in parallel processing streams describes the function of cortical pathways for information processing.

Levels of analysis describe the notion of computation in the nervous system (Marr, 1982). The computational level characterizes the conceptual tasks of problem solving, specifically how the problem decomposes into specific tasks. The algorithm level depicts the formal procedure for task performance such that, given a specific input, the correct output is produced. The physical implementation level refers to the physical device whose causal interactions implement computational principles.

As a biological machine, the nervous system performs neurocomputation based on functional mapping. The computational function of the nervous system consists of the functional mapping of elements in the sets of a physical system. The sensory input from the environment is coded into a form of the physical system or a physical state of the physical system and the physical system performs a state transition. The output from the physical system is decoded for production of the functional mapping.

The coding of sensory input from the environment into a form of the nervous system is a functional specialization of sensory systems. The nervous system demonstrates multiple methods for the encoding of featural information into physical states of the nervous system (Churchland & Sejnowski, 1992). Local coding refers to the specialization of units for detection of features distinguishable to the nervous system. Scalar coding refers to the coding of features through the firing rate of a single neuron. Vector coding refers to the encoding of features into patterns of activity from a population of units with tuning curves. The distribution of representations in the brain demonstrates the coding of featural information across units of the nervous system.

The physical properties of the nervous system demonstrate emergentism across spatiotemporal scales. The physical states of nervous systems undergo physical state transitions as causal interactions of spatial and temporal properties. The physical properties of the nervous system are emergent as physical realizers for the performance of specialized functions. The physical states of the nervous system consist of its representational content and the physical state transitions of the nervous system perform computations.

The functional specialization of the nervous system illustrates the gradual adaptation of the nervous system for specific computational purpose. The functional specialization of brain regions shows the tuning of neural circuitry for functional performance, rather than for the flexibility of computational production per se. The energy efficiency of computation of the nervous system contributes to the performance power for functional operations. The computation of the nervous system demonstrates multiple constraints along spatiotemporal dimensions.

The brain demonstrates parallel architecture for functional performance. Numerous parallel streams of information processing in neuroanatomical pathways demonstrate specialization of functional performance. The causal interactions of neural activation across neuroanatomical pathways contribute to the performance of specific functional tasks.

Neurocomputation in the real world illustrates the complexity of physical systems. The inputs from the real world are multidimensional and present a scaling problem for the nervous system. The scalar transformation of real-world inputs into neurocomputational units is required for information processing in neuroanatomical pathways. Real-world inputs consist of information that requires transformation and interpretation for real-world output. The transformation and interpretation of real-world input in bidirectional networks with hidden layers demonstrate the information processing of network activation for real-world output.

The causal relations of neural networks demonstrate the interconnected pathways of information processing for functional performance. Patterns of neural activation from a specific brain region with projections to other interconnected brain regions in the neural network demonstrate the importance of network activation for functional processing. The interconnection of brain regions within a neural network may function to produce inhibition of information processing through bidirectional projection.

Computation and Machine Functionalism

Across physical realizers, the computational components of machine functionalism consist of the set of functional tasks and their subcomponents. For simple machines, the set of functional tasks depict input–output relations given a set of rules for functional operation. For complex machines, the set of functional tasks performed require physical implementation that observes computational principles. The computational principles of biological machines illustrate the bio-physical mechanisms of computation and their causal interactions.

The computational principle of machine functionalism for simple machines is the rule-based paradigm. Rule-based paradigms refer to physical devices that use rules to perform functional operations of transformation for the production of input into output. The functional operations of rule-based paradigms consist of a machine table that describes the rules for the production of output given specific input. The rule-based paradigm shows the flexibility of simple machines that are capable of programmable input–output transformations.

Machine functionalism of the biological machine observes computational principles of the structural and dynamic aspects of the nervous system. Computational principles of the nervous system describe the fundamentals of neurocomputation in the processing pathways of the brain (O'Reilly & Munakata, 2000). Fundamentals of neurocomputation in the organization of the nervous system detail the information-processing mechanisms in its structural components. Fundamentals of neurocomputation in the dynamics of the nervous system reflect the patterns of information processing in the pathways of neural networks.

Structural principles of computation refer to the fundamentals of the organization of the nervous system. The nervous system demonstrates a hierarchical structure in pathways of processing. Pathways of processing consist of layers of interconnected networks that perform transformations of sensory input into motor output or interpretations of sensory input as internal states of network activation. Processing pathways show functional specialization in the sequence of transformations across hierarchical layers of processing. Large-scale distribution of representation refers to the information processing across large areas of interconnected pathways. Multiple pathways of processing contribute to the representational content of sensory input. Neural computation in processing pathways demonstrates dedicated, content-specific processing. Activation patterns consist of transformations for specific content defined by the stimulus. Content-specific processing in hierarchical layers of processing demonstrates how patterns of neural activation build content-specific associations across time and representations.

Dynamic principles of computation refer to the fundamentals of patterns of activity in neural networks. Multiple constraint satisfaction is a principle of brain dynamics in bidirectional interconnected networks. Attractor dynamics guide information processing to a local attractor state. The local attractor state of network activation is a harmony state. Pattern completion in the lateral processing of bidirectional interconnected networks consists of patterns of activation that contribute to the completion of activation patterns. Bootstrapping refers to the amplification of initial activation to activation patterns throughout the network.

Top-down amplification refers to the top-down amplification of bottom-up input in patterns of network activation.

The computational principles of the nervous system demonstrate the scalar complexity of the biological machine. The biological machine demonstrates input–output transformation across spatiotemporal scales. Neurocomputation of the biological machine requires multiple pathways of information-processing mechanisms to perform the production of a motor output or response that is functionally appropriate given the stimulus or sensory input. Neurocomputation consists of the completion of patterns of activation in neural networks that satisfies multiple constraints.

Machine functionalism demonstrates the importance of the functional role of the mind in the physical implementation of mental computation. Because mental computation has multiple physical realizers, the functional role of a specific mental task is the fundamental element of the mental computation. The computation of physical realizers performed in a functional task that is defined by the stimulus or specific to sensory input demonstrates functional specialization or the content-specificity of computation. The functional specialization of computational mechanisms shows adaptation of mental content to the environment across spatiotemporal scales. The transformation of sensory input into real-world output depicts the functional role of computation for real-world advantages.

Culture and Machine Functionalism

Culture consists of the shared meaning systems that guide mental processes and their underlying mechanisms for the production of behavioral adaptation. Culture refers to the set of functional tasks and its components in the mind for the production of behavioral adaptation. Culture consists of the mental content and its causal roles and relations for the transmission of cultural information. The cultural patterns of the mind comprise shared meaning systems that shape the production of behavioral adaptation.

Machine functionalism contributes to philosophical investigations into the nature of the mind and its function. In machine functionalism, the cultural mind consists of cultural mental property. Cultural mental property connotes the set of cultural mental events as part of a causal system. The cultural mind as a causal system refers to the network of cultural mental events as a causal power. The causal power of cultural mental events demonstrates the functional performance of the mind.

The cultural mind as a cultural network of mental events is comprised of mental content with causal roles and relations. The cultural property of the mind is a multilayered network of cultural mental states that perform specific functional roles and cultural patterns of causal relations. In the cultural network, internal states of representational content reflect the interpretation of cultural sense data or the transformation of cultural sense data for cultural production. Cultural mental content is the internal states of representational content in cultural thought. Cultural mental content comprises the cultural pattern of internal states of the cultural network. The cultural patterns of thought refer to cultural mental content and its causal relations.

The notion of the cultural mind as cultural mental property posits the notion of the mind as physical property in a physical system. The multiple realizations of the cultural mind as cultural physical property imply the importance of the functional role of cultural mental property and its causal relations across physical instantiations. Cultural mental states as cultural physical states consist of spatial and temporal properties to show the capability of functional performance in a cultural physical system.

Culture as a causal system consists of the mental and physical property of culture. The mental and physical property of culture is comprised of the mental and physical states of culture and their spatiotemporal properties. Cultural dynamics refer to the rate of change in spatiotemporal properties of the cultural system. Cultural property in mental and physical states depicts a configuration of parts as a complex set of causal effects within the physical system.

Cultural computation refers to the implementation of functional tasks and its components. Early notions of the cultural mind in machine functionalism are a conceptualization of the mind as a cultural computing machine. The cultural mind as a cultural computing machine performs specific functions. Functional performance of the cultural computing machine is the production of cultural output from cultural input from culture-based rule paradigms. Culture-based rule paradigms are specifications of cultural rules for machine production given a particular input.

The cultural machine as a deterministic machine demonstrates the capability for the machine production of cultural content. The cultural computing machine is capable of producing specific predetermined cultural content given a particular input and machine table. The cultural machine table defines rules of cultural computation for machine production. The cultural machine is programmable as a universal machine for the deterministic machine production of cultural content. The deterministic production of cultural content from the cultural machine contributes to the regularities in cultural patterns.

Cultural probabilistic automatons show the capability for probabilistic machine production of cultural content. Cultural probabilistic automatons demonstrate that machine production is not inherent to a particular input, but instead programmable through rule-based paradigms. The cultural machine as a probabilistic automaton produces cultural content based on probabilistic machine state transitions. Cultural input of a causal output that is probabilistic contains the cultural mental property or cultural mental events that are required for a fixed rate of a given outcome. Hence, the cultural computing machine shows the capability for the machine production of cultural content that is probabilistic given conditions of uncertainty.

The notion of probabilistic cultural automatons suggests the relevance of strategies of prevention and intervention in machine state production and their respective machine state transitions. Prevention strategies refer to the functional operation of the rule-based paradigm to ensure the machine production of no probability for a particular output. Intervention strategies refer to the functional operation of the rule-based paradigm for the machine production of no probability of a particular output in a machine state transition. The use of prevention and intervention strategies in probabilistic cultural automatons demonstrates the programming of the cultural machine for an optimal state of function.

The cultural mind as a biological machine illustrates the physical instantiation of cultural processes in the natural world. The cultural brain is an emergent pattern of natural phenomena. The cultural brain demonstrates the biological plausibility of computational models. The cultural brain consists of the functional mechanisms and their causal interactions that generate cultural patterns of thought as emergent properties. The cultural brain demonstrates cultural computation in the natural world. The cultural neurocomputation of the brain illustrates the causal explanations of culture as patterns of activity of the nervous system. For biological organisms, culture processes are comprised of the set of mental tasks that perform an adaptive function. The mental property of the cultural mind as a biological machine consists of the mental content that ensures cultural and psychological adaptation of the organism.

At the individual level, cultural computation shows functional specialization through information processing that is content-specific and dedicated to the cultural stimulus. The physical instantiation of cultural mental events in the brain demonstrates the causal power of cultural computation in the biological organism. The brain produces the mental property of the cultural computing machine. The brain shows the capacity to produce the mental events that are causal events of the cultural machine. The mental property of the cultural machine as a biological organism is a causal power.

At the group level, cultural computation demonstrates the implementation of functional roles in cultural and social patterns of the mind and machine. Cultural and social patterns of thought consist of social norms that perform specific functions for cultural and psychological adaptation. The cultural mind as a cultural computing machine can perform cultural output based on specific rules defined by social roles. Culture-based social rule paradigms are descriptions of cultural rules for the transformation of social input to social output. The cultural machine demonstrates the construction of culture from social patterns of thought.

Cultural neurocomputation illustrates the physical instantiation of the cultural mind in the biological organism. Cultural neurocomputation refers to the cultural processes in the structural and functional organization of the nervous system. Cultural neurocomputation across levels of processing of the nervous system illustrates the functional architecture of cultural information-processing mechanisms. Cultural information-processing mechanisms encode and transform cultural sense data into the production of patterns for cultural adaptation. The cultural neurocomputation of biological organisms shows the transformation and interpretation of real-world input for real-world advantage.

Cultural functional equivalence in the level of performance of a cultural machine and a cultural brain illustrates the importance of functional role in the cultural mind. The input–output equivalence of a cultural machine and a cultural brain does not necessarily constitute the same level of cultural performance. The rule-based paradigm of cultural computation illustrates the importance of distinct cultural identity for mental computation.

Cultural computation in the biological machine contributes to the robust patterns of mechanisms and their causal relations of the cultural brain. The functional relations of the cultural brain, its cultural mental and physical states, describe cultural patterns in nature that hold. Cultural processes of the nervous system demonstrate the lawlike regularity of cultural mental life. Cultural mental life as a causal power refers to the mental content that contributes to the regularities of cultural patterns.

Culture, Minds, and Machines

Cultural processes are the shared meaning system comprising values, practice, and beliefs of people defined by ancestry, language, geographic origin, customs, and ethnic heritage. Cultural processes consist of the property of the cultural group that performs a functional role in the real world and are defined with spatio-temporal properties. Cultural dimensions guide the knowledge representation across the sharing of the mental states of self and others.

In the cultural machine, the cultural machine table defines specific rules for the production of particular cultural output. The rules specified in the cultural machine table describe specific configurations of machine states and its transitions based on deterministic cultural input–output mappings. The cultural machine can be programmed with culture-based machine tables to perform any specific output.

The cultural mind as a cultural computing machine consists of the mental functions that generate cultural mental content. The cultural mental content of the mind consists of the mental and physical property of the mind at the cultural level. The cultural mental property and cultural physical property of the cultural mind exist as a functional mapping of a set of identity relations. The mental and physical property of the cultural mind consists of the set of mental states and mental events at the cultural level. The physical instantiation of cultural mental property is the realization of its cultural mental content across multiple physical realizers.

The cultural mental states play a causal role among multiple physical realizers. Cultural mental states as machine production show how cultural processes are instantiated as a set of physical states and their physical state transitions. Cultural mental states as the set of physical states from biophysical mechanisms of the biological organism consist of cultural mental content as emergent property. The emergent property of the cultural brain is comprised of the cultural mental states as mental events of real-world experience. The emergent property of the cultural level as an organized system consists of the parts and particulars of cultural patterns and regularities in the total causal structure in the world. Emergent property as cultural property has causal power in discovery processes as explanatory inference and prediction.

The cultural mind as a performance of causal functions is a causal power in the world. Emergent phenomena as cultural mental content consist of the discovery process from explanatory inference for truth correspondence of the mind in the world. Emergent phenomena as cultural mental content describe the intentionality of the mind and its causal-functional role at the cultural level. Cultural mental content as real-world experience in the physical world is the emergent property of the cultural brain as part of the physical system. The cultural mind as states of consciousness contributes to the cultural mental content from the experiential interaction of the mind in the world.

The notion of the functional equivalence of the cultural mind and the cultural machine implies a commonality in causal-functional roles. The functional equivalence of cultural content from the cultural mind and the cultural machine suggests a commonality of input–output relations. The cultural mind as a causal power is the causal source of cultural mental content in the world. The cultural mind plays a causal-functional role in the production of cultural mental content. The cultural machine demonstrates automated production based on cultural mental content, including the input–output relations and culture-based rule sets.

The multiple physical realizations of the cultural mind are a causal power in the structure of the world. The mental content of the cultural mind and cultural machine that are identical suggests a sameness in the physical realizations of the mind. The mental content of the cultural mind and cultural machine that are identical implies a likeness in the causal power of the cultural mind in the causal structure of the world. The mental content of the cultural mind in the cultural machine entails the causal role of truth theories in correspondence to the causal structure of the world. The importance of the distinction of the mental content of the mind and machine that is apparent at the individual level is epiphenomenal at the cultural level.

The causal power of the cultural mental content of the cultural machine is relational to its functional operations. The cultural mental content of the cultural mind plays a causal-functional role in the functional operations of the cultural machine. The mental content of the cultural mind performs so as to command the content of the cultural machine. The cultural mind performs the functional operations of the cultural machine as its superintention. The mental content of the cultural mind supervenes as the mental content of the cultural machine. The performance of cultural mental content from the cultural machine is a slow simulation of the cultural mental content of the cultural mind. The cultural mental content of the cultural machine performs to share in sameness and likeness with the causal power of the cultural mind.

Conclusion

Machine functionalism is one of the early conceptualizations of the mind as a machine. The mind as a machine metaphor has branched into several areas of core considerations in philosophy of mind. The mind as a machine broadens the conceptual spectrum of the functional relation of the mind and the brain. The functional relation of the mind as the brain has furthered interest in the understanding of the mind across multiple physical realizations.

The cultural level of the mind in the world advances the metaphor of the mind as a machine in novel directions. The cultural mind implies the causal-functional role of the cultural machine. The cultural brain as a cultural computer entails the features of the biological organism that comprise the parts and particulars of the cultural level of the physical system. The notion of the cultural mind as a cultural machine introduces novel considerations of the causal-functional role of intentionality. Understanding the mind as a machine metaphor at the cultural level expands the realm of inquiry into the cultural mind as a causal system. The broad understanding of culture in minds and machines contributes insight into the fundamental nature of the mind in the world.

References

Churchland, P.S. & Sejnowski, T.J. (1992). *The computational brain.* Cambridge, MA: MIT Press.
Marr, D. (1982). *Vision.* New York: Freeman.
Norenzayan, A. & Heine, S.J. (2005). Psychological universals: what are they and how can we know? *Psychological Bulletin, 131,* 763–784.
O'Reilly, R.C. & Munakata, Y. (2000). *Computational explorations in cognitive neuroscience: Understanding the mind by simulating the brain.* Cambridge, MA: MIT Press.
Turing, A. (1950). Computing machinery and intelligence. *Mind, 59,* 433–460.

Further reading

Kim, J. (2011). *Philosophy of mind.* Boulder, CO: Westview Press.

PART II

5
RECONSTRUCTIONISM

Introduction

Reconstructionism represents an empirical approach to the study of the mind, brain, and behavior that emphasizes the processes of basic mechanisms of the mind and brain and their emergent phenomena. Reconstructionism as a philosophical position articulates that the component processes of the mind and brain and their interactions describe complex and emergent phenomena. Emergent phenomena refer to the production of mechanisms and processes that are cumulative and arise specifically through interactions of the basic mechanisms and processes. Emergent phenomena are representative of basic mechanisms and processes that are produced in interaction with each other. Emergent phenomena are sets of relational phenomena as properties that occur in interaction with each other.

In philosophy of mind, reconstructionism represents a relational approach to the understanding of the mind from the brain. Reconstructionism is a philosophical position similar to emergentism that explains mental phenomena as a product of neurobiological mechanisms (Kim, 2011). Reconstructionism assumes that mental phenomena are emergent products of the interaction of identity relations from sets of mental and neural properties. Mental phenomena are the emergent properties of the interaction of relational occurrence of mental and neural properties as mental and physical events. The conceptualization of mental phenomena as a product from an organized system of aggregate parts implies that it arises as an emergent part of the reliable patterns and regularities of mental thought in the natural world.

Reconstructionism assumes that the mind, brain, and behavior cannot be understood merely through reductionism of these mechanisms as component parts. Rather, reconstructionism posits that the formal process to build a given phenomenon from component parts allows for the discovery of the levels of organization of a given system, such that the layers of computational mechanisms and processes that produce the phenomena are revealed (O'Reilly & Munakata, 2000). Reconstructionism is a philosophical notion for the discovery of the inter-activity of processes and mechanisms as emergent phenomena.

In philosophy of science, reconstructionism is a complementary process to the complete explanation in the physical world. Reconstructionism implies that the interaction of the mind in the world leads to the production of reliable patterns and regularities in the natural world from mentality and reasoning. Reconstructionism is the amplification of the importance of the set of identity relations of mental thought and neural mechanisms as emergent property in the parts and particulars of the physical system as an organized system. Reconstructionism entails the identity relations of mental thought and neural mechanisms as emergent property that contribute to the parts and particulars of the total causal structure in the world.

Reconstructionism is a discovery approach to a complementary and satisfactory explanation of the production of a given phenomenon. Reconstructionism as a discovery model of emergent phenomena can produce novel predictions. Computer programs that perform simulations of emergent phenomena are designed to test complex phenomena. Computational models serve as a formal test of hypotheses that produce the emergent phenomena. Reconstructionism consists of discovery models that assume that the production of emergent phenomena arises from an organized system of parts. For processes of discovery that are independent of verbal argument, reconstructionism provides a scientific approach to the determination of the emergence of natural phenomena.

The processes of reconstructionism contribute to computational approaches in cultural neuroscience, such that simulations of theoretical models are built and tested to formalize the characterization of the mechanisms of cultural and neurobiological systems. Computational models of cultural neural networks reflect the notion that the formal characterization of mechanisms of cultural and neurobiological systems depicts pathways of network activity. Computational models of cultural neural networks test the biological plausibility of cultural inferences from structural and functional organization of the nervous system across levels of analysis.

The graphical schemata of basic mechanisms and their emergent phenomena in a computational model allow for characteristics of such processes to be explicitly represented of a given mental phenomenon. The computational model of thought reflects a way of building a formal theory regarding the antecedents and consequences of different constructs. Computational approaches in cultural neuroscience provide a unified framework with which hypothetical interactions and behaviors of a given system can be theorized and tested in a systematic manner.

Understanding mental phenomena as emergent phenomena contributes to the truth correspondence in the causal structure of the world. The philosophical postulation of mental phenomena as emergent phenomena is the understanding that the production of mental content as explanation is because it is part of the truth theories in the structure of the world. Emergent mental content as explanatory inference is the correspondence of mental content with such truth theories. The emergence of mental content as explanatory inference demonstrates the causal power of the discovery approaches of reconstructionism.

Emergent phenomena as the mental content from prediction entail the production of mental inferences from patterns of thought. Mental inference as prediction is the amplification of part of the truth theories in the structure of the world in the mind in interaction with a physical state transition in the world. Mental inference as prediction is emergent phenomena in the mind as a causal power in the world. Emergent mental phenomena as prediction acts as a causal power with the intentionality for truth correspondence in the mind and in the world.

The states of consciousness as emergent phenomena describe knowledge generation from the experience of first-person perspective. The stream of subjective experience as the mind in interaction with the world is mental phenomena as emergent phenomena. Conscious awareness of the mind in interaction with the world is the mental and physical property as part of the organized system. Mental awareness as states of consciousness is causal power with the intentionality for truth correspondence with the structure of the world.

Computation and Reconstructionism

Reconstructionism is the emergent parts and particulars of organized systems as the underlying mechanisms of natural phenomena in the structure of the world. Reconstructionism consists of the emergent phenomena as a component of the complete parts and particulars of the organized system as a living system. In reconstructionism, the interaction of emergent parts is greater than the independence of such parts. The interactivity of emergent parts contributes to the larger patterns and regularities in the structure of the world. Reconstructionism implies that the interactivity of emergent parts into a synthesis is a source of the performance and production of natural phenomena. Reconstructionism entails that emergent parts act as a causal mechanism in a complex set of effects within the organized system.

Reconstructionist accounts of thought and reason posit that mental thought and reason act as emergent parts that have a causal power. The causal power of mental content implies that the higher-level features arise from bottom-up processes as emergent parts. Reconstructionist notions imply that the patterns from the interaction of emergent parts are independent to an extent within the organized system. In reconstructionism, the patterns of emergent parts are a standard of criteria in the explanation and causation of natural phenomena.

Emergent mental phenomena describe the causation of knowledge generation in terms of the physical implementation of neural mechanisms. Mental causation emerges from the patterns of mental thought and neural activation. Patterns of neural activation act as a causal mechanism of mental thought. Mental thought is considered to emerge from patterns of neural activation. For emergent mental phenomena, mental thought from neural activation acts as a causal mechanism in the physical world. Knowledge generation as emergent mental phenomena is a causal power for reasoning about the structure of the world.

Reconstruction describes the flow of information from bottom–up processes. In computational simulations of neural networks, bottom–up processes consist of an organized system of parts as mechanisms. The structural and functional organization of the nervous system consists of levels of processing and levels of analysis. Neural networks are a level of information-processing mechanism that refer to the interactivity of the flow of information as the connectivity across the network. From representation to the transformation of input into response, bottom–up processes are the information flow as network activation patterns of interactivity to an optimal state or a final state of activity. For knowledge generation, bottom–up processes entail the physical and mental processes for the transformation of sensation into the production of mental thought.

The structural and functional organization of the nervous system describes the biophysical mechanisms that comprise levels of processing. Mental content and its representation are encoded and transformed across levels of processing. From molecular and cellular mechanisms to neural systems, the levels of processing of the nervous system detail the functional architecture of the brain across spatio-temporal dimensions. The principles of structural and functional organization of the nervous system depict the flow of information processing across levels as the encoding of sensation into units of information representation and its transformation into response.

The interaction of parts of the organized system is greater than each of the parts independently. For bottom–up processes, information flow in the feedforward network interacts with the network through unidirectional connectivity. Feedforward networks can perform transformations of input to output as distributed and localist representations. The transformations of such representations comprise the emphasis on certain distinctions. Distributed representations act as a unit that represents multiple inputs; similarly, each input is represented across multiple units. Localist representations include representations of input in a single unit.

As information flows forward in the network from unidirectional to bidirectional connectivity, the directionality of information flow in the network activation is greater in the unidirectional net independently. The bidirectional network allows information flow to feed back from input to output units across hidden layers as in a recurrent net. The feedback of information flow across hidden layers contributes to inhibitory interactions that control levels of activation. Inhibitory interactivity provides a mechanism of selection.

Reconstructionism as a bottom–up process is complementary to the top-down processes of reductionism. Reductionism in the physical system consists of the parts and particulars of the underlying mechanisms of natural phenomena in the world. Reductionism entails that mental phenomena as a concept are a top-down influence on the parts and the particulars of the organized system. By contrast, reconstructionism emphasizes the bottom-up processes for the production of mental thought from physical implementation that contribute to the larger interactions of parts of the organized system.

Culture and Reconstructionism

Reconstructionism is a computational approach to the study of how the individual generates cultural inferences from mental and neural states and how cultural systems regulate the mental and neural states of the individual as emergent phenomena. Reconstructionism is a computational approach to the studying of the processes of cultural transmission of mental and neural states across the group of individuals as parts of an organized system. Computational modeling of cultural processes allows for the formal testing of theoretical frameworks of how culture affects the mental and neural states of the group and individual. Different factors affect the characteristics of the computational model of cultural processes. In general, computational modeling of cultural systems consists of the simulation and construction of the computational components of culture.

Cultural niche construction is a theoretical model that describes computational components of biologically plausible living systems at the cultural level. At the level of the group, cultural niche construction characterizes the processes such that organisms are part or components of an organized system and construct a cultural system through their changes in the ecological niche. At the level of the individual, components of the nervous system interact to produce cognitive processes of the cultural system.

A reconstructionist theory of culture conceptualizes cultural thought as an emergent phenomenon from interactions of mental states and neural mechanisms. In a reconstructionist theory of culture, reconstructionism describes the production of culture as emergent phenomena from the interaction of component parts. Cultural inferences arise from the interactivity of informational processing mechanisms with the environment. Culture as knowledge generation is the production of cultural thought from the developmental maturation of the functional architecture of the brain. At the level of the organism and ecological niche, the interaction of the mind in the world results in cultural production that is emergent and uniquely characteristic of their interaction.

A reconstructionist theory of culture positions cultural phenomena as emergent property from the interactivity of component parts. Cultural phenomena are emergent property that holds the components of causal interaction. Cultural mental phenomena as emergent property hold the causal explanation of cultural mental content. Culture as emergent property is a source of supervenience of mental property to physical property. Culture as emergent property is a source of the creation and maintenance of cultural property, the mental and physical property of minds and machines at the cultural level.

Cultural systems provide boundaries for physical and social interactions among group members that vary in size, frequency, breadth, and depth of social communication. Cultural systems refer to the strict or broad adherence to societal norms. Cultural dimensions describe expectations of group members to adhere to societal norms during social interactions. Cultural dimensions provide social rules for social interactions of members in groups that vary in size and frequency of social communication. Computational modeling of culture shows how the network of social interaction of group members differentiates between strict and broad adherence to societal norms. Behaviors in social interactions that conform to societal rules within a given context are predicted based on the societal norms of cultures.

Computational modeling of culture is consistent with the dynamic principles of multiple constraint satisfaction. The activation state of behaviors is based on the satisfaction of constraints from environmental inputs and learned weights. Environmental inputs that feed forward into the cultural net refer to natural and man-made ecological threats. Learned weights in the cultural net reflect system-wide responses to environmental inputs, such as responses within mental and neural systems of behavior. Computational modeling of culture may be characterized as cultural learning models, or cultural algorithms of error-driven learning. Behaviors in social interactions that are different from those expected given the societal rules of a given context may be considered manifestations of cultural learning.

Cultural systems consist of different physical states that vary in levels of energy. A physical system of culture can be described with an energy function that describes a global measure of energy of the system based on a system feature and the strength of the interactions between members in the system. The energy level of a cultural state may be considered lower than that of another cultural state, due to the greater satisfaction of constraints within the spatiotemporal dimensions of the cultural system.

Within the individual, a physical system of culture can be described with an energy function that describes a global measure of energy of the system based on neural activation and the strength of the interactions between neurons in the neural system. The cultural brain dynamics may be described as having lower energy and greater constraint satisfaction compared with the brain dynamics in a different cultural dimension. Cultural dynamics consist of brain dynamics with higher energy and lower constraint satisfaction given environmental inputs and learned weights. Brain dynamics in cultural dimensions may reflect the responses in learned weights to environmental inputs.

The computational principle of reconstructionism describes basic mechanisms that contribute to the development and implementation of cultural systems and their regulation in the environment. Cultural systems may be designed to produce states of rationale through the interaction of component parts of natural and artificial systems. Novel mental and physical states of reasoning may be produced for individuals that are designed in cultural systems which are unique. The production of mental and physical states from the cultural system may also emerge from the development and implementation of cultural systems designed with the reconstructionism principle. In natural systems, culture as information production describes knowledge generation to inform decision making and governance.

The emergent phenomena of the cultural system may consist of the inter-activity of mental and physical states of individuals in the environment that are distinct processes. The regulation of cultural systems emerges from the weighted interaction of the component parts. The distinct processes of emergent mental and physical phenomena unfold through the interaction of multilevel mechanisms. The emergent mental phenomena of cultural systems are produced for the cultural construction or reconstruction of culture.

Reconstructionism in Culture

A cultural theory of reconstructionism postulates that the generation of cultural mental content is the emergence of mental phenomena as part of the organized system at the cultural system level. Cultural mental content as emergent phenomena consists of the parts of the organized system. Cultural patterns of thought emerge from the local brain dynamics of minds in interaction with the world. Cultural patterns of thought arise from the knowledge generation of the mind. Cultural mental content consists of the explanatory inferences and causal reason that emerge from the interaction of the cultural mind in the world.

Cultural mental content as a stream of conscious experience is the cultural experience of the first-person perspective. Cultural experience as the first-person perspective implies that the stream of consciousness that is private, personal, and self-determined. Cultural experience as the generation of knowledge for mental performance is the function of the first-person perspective. Conscious experience as cultural mental content reflects the causal role of states of consciousness as emergent phenomena that are part of the cultural level of living systems.

Cultural knowledge as first-person knowledge is the generation of knowledge for mental causation in the organized system at the cultural system level. Conscious experience as cultural experience in first-person knowledge is the awareness of the mind of the causal power of cultural mental content and its interaction with the structure of the world. Conscious experience as cultural experience is the states of consciousness that exist as emergent phenomena from the intentionality of truth correspondence with the structure of the world.

Cultural mental content as the generation of knowledge for information production is the cultural experience of the third-person perspective. Cultural mental content as emergent mental property for information production is the function of the third-person perspective. Cultural experience as the third-person perspective implies that the stream of consciousness is public, shared, and deterministic.

Cultural mental content as the cultural experience of the third-person perspective implies a stream of conscious experience that is part of the structure of the world.

Cultural mental content as third-person knowledge is the states of consciousness as emergent phenomena in truth correspondence with the structure of the world. Cultural content as third-person knowledge entails the generation of knowledge that is consistent with parts of the structure of the world. Cultural mental content as third-person knowledge comprises the truth theories that are deterministic and hold regularities in the patterns of the natural world. Cultural knowledge as third-person knowledge is the generation of knowledge for the creation and maintenance of cultural tradition.

Cultural mental content as conscious experience is the amplification of the set of identity relations of cultural mental thought and cultural neural mechanisms as the parts and particulars of the physical system. Cultural consciousness entails a discovery model of emergent phenomena and knowledge generation for the prediction of natural phenomena. Cultural consciousness is a discovery process to a complete and satisfactory explanation of natural phenomena. Cultural consciousness is a discovery process of the truth theories in the structure of the world. Culture as a stream of consciousness is the mind in experiential interactivity with the world.

The cultural system level consists of the performance of mental and machine computation for information production. Mental computation as internal information production at the cultural system level consists of cultural patterns of thought. Cultural patterns of thought demonstrate the computational principle of reconstructionism in the mental computation of the cultural system. Information-processing mechanisms in the structural and functional organization of the nervous system comprise the computational components necessary for cultural patterns of thought. The biophysical mechanisms of the nervous system as the functional operations of information mechanisms generate the mental computation for cultural patterns of thought across levels of processing.

The cooccurrence of cultural mental thought and neural information-processing mechanisms describes the cultural patterns of thought that guide societal perception. The generation of cultural patterns of thought contributes to the maintenance of cultural tradition and cultural identity. Cultural patterns of thought comprise the regularities that define the emergent phenomena at the cultural level. The generation of cultural patterns of thought demonstrates the causal power of the mind to guide societal perception. The generation of cultural patterns of thought illustrates the causal role of the mind to produce the cultural mental content that creates and maintains cultural traditions.

Knowledge generation as cultural patterns of thought guides cultural transmission of beliefs, values, and practices. Cultural transmission of mental content demonstrates the persistence of cultural mental content from the cultural model to cultural learners. Computational models of cultural transmission describe the simulation of cultural patterns of thought that are acquired from model to learner.

Cultural transmission from knowledge generation holds differential rates of transmission based on the cultural mental content.

The emergence of cultural mental content in the mind entails the discovery of cultural processes in the biological machine. Cultural phenomena as emergent mental phenomena consist of the performance of mental constructs for the production of cultural content. Emergent cultural phenomena as a cultural performance are the function of mental imagination for cultural inference. Cultural inference acts as a generator of novel prediction for the construction of the cultural niche. The emergent cultural phenomena of the mind generate the explanatory inferences and causal predictions necessary for cultural construction in truth correspondence with the world.

The cultural content of the mind in the discovery of cultural processes centers on the emergent phenomena of mental content in the natural world. The conception of thought and reason as emergent phenomena is the creation of mental content at the cultural level. The concept of thought defines a standard of criteria for the emergence of mental phenomena. The creation of mental content entails the emergence of mental phenomena as patterns in the natural world. The creation of mental content describes cultural patterns of thought in nature that hold. The parts and particulars of cultural mental phenomena are consistent with those spatiotemporal properties in the natural world.

Mental imagination as emergent cultural phenomena implies the states of conscious awareness of the cultural content of possible worlds. Imagination as emergent mental phenomena entails the creation of cultural content for the construction of possible worlds. Mental imagination consists of the set of mental content that supervenes on physical content. Through the imagination of the mind, emergent mental phenomena are a causal power for the creation and construction of cultural content across possible worlds.

Cultural phenomena as emergent mental phenomena describe one of the robust regularities in the structure of the world. Cultural mental phenomena as emergent property provide a guide of explanatory inferences and causal predictions that guide the parts and particulars of spatiotemporal properties in the structure of the world. Cultural mental phenomena consist of the causal predictions that define the deterministic mental causation of possible events. Emergent cultural phenomena from the mind are a causal power of cultural performance and cultural production in interactivity with the structure of the world.

Cultural patterns of thought that guide societal perception illustrate the causal power of knowledge generation to a complete and satisfactory explanation. Cultural thought that guides knowledge generation acts as a correspondence of truth theories in the structure of the world. Cultural patterns of thought hold explanatory inferences and predictions that are part of the truth theories. The emergence of cultural patterns of thought in the mind demonstrates the causal power of inferential reason to create and maintain the cultural level through interactivity of the mind in the world.

Cultural content builds the cultural property of the biological machine in the world. Cultural mental content from knowledge generation of the mind contributes to the mental property of the cultural level. The cultural neural networks as neural information-processing mechanisms that cooccur with cultural mental content are part of the physical property of the cultural level. Cultural physical property of the mind consists of the functions and operations as mechanisms from the biological machine. Cultural content as mental and physical property contributes to the cumulative cultural property of the organized system.

The emergence of cultural content in the mind and in the world demonstrates the cultural capital of the biological machine. The emergent cultural property of the mind is part and particular of an organized system at the cultural system level. Emergent cultural property entails the function of adaptation of the cultural property for individuals and groups. The emergent cultural property of the mind in the world illustrates the valuation of cultural mental content from the individual to the group. Emergent mental phenomena as cultural property imply the truth value of the cultural mental content from the biological machine to the structure of the world. Cultural property in the mind and in the world contributes to the overall cultural capital of the organized system.

Emergent mental phenomena are part and particular of the cultural security of the organized system. Emergent cultural phenomena of the mind consist of the cultural mental content that performs functions of prevention and intervention of the mind in interaction with the world. Emergent cultural mental property is a generator of the mental state inferences that contribute to the cultural inferences that shape social processes and societal organization in the world. Emergent cultural mental property consists of the tacit metaphysics and epistemology that guide knowledge generation across ancient philosophical traditions. Emergent cultural mental property is the generator of culture-based causal attributions that guide societal processes of nation states. Emergent mental property and its reconstruction contribute to the production of strategies for the restoration and replication of cultural mental content. Emergent cultural mental property is the design of strategic resources of information and plans of action for the prevention and recovery from threats as part of human security.

Emergent cultural phenomena contribute to the economic and social empowerment of individuals, societies, and nations. Cultural mental phenomena as emergent property entail the processes of truth valuation of cultural content of minds and machines in the world. Cultural mental phenomena as emergent property imply the production of explanatory inference and prediction for theory confirmation of the truth correspondence of cultural content in the structure of the world. The cultural content of the mind informs the coordination of individuals and groups for cultural participation. The broadening of cultural participation of individuals and groups contributes to the coordinated efforts for cultural development. The cultural content of the mind contributes to the scientific and technological progression that leads to cultural advancement. Cultural phenomena of the mind as emergent property imply the causal power of the mind for the fulfillment of human potential.

Minds, Machines, and Reconstructionism

In philosophy of mind, reconstructionism is one of the strongest accounts of mental computation as machine computation. Reconstructionism assumes that mental and physical states as identity relations are the fundamental property of thought. Reconstructionism implies a standard of mental thought that is consistent with the fundamental property of thought as emergent property from an organized system. The philosophical notion of reconstructionism entails the emergence of thought from neural mechanisms. The states of neural mechanisms play a causal role in the generation of thought as an emergent property. Thought as an emergent property is the identity relation of mental and physical states in cooccurrence with each other.

Reconstructionism places emphasis on the importance of the formal testing of models of mental processes. Reconstructionism places into the discovery how mental thought emerges from basic mechanisms of the brain. The computational principles of reconstructionism describe the features of basic mechanisms that contribute to the generation of thought. Reconstructionism entails the concept of thought as mental computation from the biological machine.

The mind as a set of mental states arises as the emergent property of the organized system. The brain as a biological computing machine comprises a set of physical states as neural states that cooccur with mental states. The cooccurrence of mental and neural states is the identity relation as an emergent property of the organized system. The cooccurrence of mental and neural states as a pattern of thought comprises an optimal state of the mind as part of the organized system.

Thought as the emergent property from the flow of information of the biological machine consists of a sequence of physical states. The flow of information as a set of activation states in the neural networks of the biological machine consists of the transformation of sense data into response. The transformation of sense data into internal representations of the biological machine reflects the feedforward flow of information into neural networks. The internal representations of the biological machine comprise interpretations of the sense data in its state of transformation.

The internal representations of neural networks describe the weighing of units of activation based on the environmental input and the biological organism. The weighing of units of activation of the neural network can be adjusted based on the directionality of the flow of information across hidden layers of network activation. The feedforward flow of information into neural networks describes the bottom-up processes that contribute to internal representations. Internal representations of neural networks emphasize the specific features consistent with specialized activation patterns of bottom-up processes.

The basic mechanisms of bottom-up processes consist of the functional specialization of neurons for the detection of specific features. The functional specialization of neurons implies the detection of specific features from sense data for the production of a specific response. The transformation of sense data involves the emphasis of specific features to produce activation patterns that are similar to those of basic mechanisms. The similarity structure of internal representations comprises the basis for the transformation of sense data.

Reconstructionism is a primary philosophical account of the production of emergent property from interactivity, whereby the interaction of smaller components produces emergent phenomena that are larger than the individual parts. The organizational and functional structure of the nervous system produces complex mental phenomena from the interaction of smaller component parts. The emergent construction of mental phenomena from the interaction of computational components highlights the importance of the interactivity of physical states for the production of emergent property. Reconstructionism defines the notion of emergent property as a production of components in interaction. The product of the interaction of components is itself the emergent property.

In a broad sense, reconstructionism as a philosophical notion ensures that there exists emergent property or the emergence of property from the interaction of parts that is larger than the components. The importance of the functional and causal roles of mental and physical identity relations as a standard of the causal conception of mental thought is apparent from earlier philosophical accounts of mental functionalism. Reconstructionism introduces the causal conception of interactivity as a standard that defines the functional and causal roles of emergent property. The interactivity of component parts is the performance of a functional role for the production of emergent property. The interactivity of component parts is the causal power that produces emergent property. Thus, the functional and causal roles of interactivity are a standard of criteria that defines emergent property.

Reconstructionism implies that the emergent property from the interactivity of mental and physical property is the product of change in the spatiotemporal dimensions of the world. Emergent property as the product of change in spatiotemporal properties implies the expansion of spatiotemporal dimensions based on component parts. The expansion of spatiotemporal dimensions for the creation of emergent property is a mechanism for the building of mental capital. The generation of the emergent property of minds is the building of mental capital to the fulfillment of human potential. Knowledge generation of the mind builds mental capital through the actuality of mental property as emergent property.

The actuality of mental property as emergent property is knowledge generation that is in truth correspondence with the structure of the world. The actualization of mental property assumes the existence of the emergent property in the world. The physical realization of mental property implies the actualization of mental property in the mind and the potentiation of its existence in the world through discovery. The actuality of emergent mental property is an assurance from the production of mental thought in the interactivity of the mind. The actuality of mental thought is the confirmation of facts through scientific observation and explanatory inference.

In philosophy of science, reconstructionism is a source of the satisfactory and complete explanation from discovery. The emergence of mental property implies that interactivity is a source of the satisfactory and complete explanation. Theories of unification of a large range of phenomena are thought to hold theories of "explanatory promise." Emergence as natural phenomena consists of the property that holds the causal history of the event, or a good or complete explanation of the emergent natural phenomena in the world. Component parts of emergent phenomena in the natural world are the causal parts of the good or complete explanation. The discovery of emergent phenomena in the natural world holds the pieces of the satisfactory and complete explanation.

Reconstruction as the simulation of mental thought from computer program is the production of machine computation as mental computation. The production of mental content from machine computation is a physical realization of mental thought from the interaction of parts of the computer. The machine computation of mental content is a simulation of the biological machine for the production of machine property. The production of machine property is a resource for the development and implementation of programs for societal benefit. The production of machine property at the cultural level contributes to the valuation of machine capital. In reconstructionism, the simulation of the biological machine is a standard of criteria for the production of machine property. Machine property is the verification of the mental content of emergent property from its reconstruction.

Conclusion

Reconstructionism is a fundamental component of thought. Reconstructionism highlights the processes of the construction of thought from component parts. The capacity of neural networks to generate specific patterns of thought as states of consciousness demonstrates the functional role of mental computation as emergent parts of mental phenomena in the natural world.

Reconstructionism is a philosophical approach similar to emergentism. In philosophy of mind, emergent property as mental and physical property from parts has a functional and causal role in the physical world. At the cultural level of the physical system, a reconstructionist theory of culture entails that culture is emergent property from mental and neural states. A cultural theory of reconstructionism posits that culture is emergent property from interactivity of minds as parts of the organized system at the cultural level.

Reconstructionism is a strong stance in support of interactivity as a mechanism of emergent property. Emergent property is a cooccurrence with the expansion of spatiotemporal dimensions in the physical world. Emergent property is an assurance of the actuality of thought and underlying mechanisms as regularities and patterns in the structure of the world.

References

Kim, J. (2011). *Philosophy of mind*. Boulder, CO: Westview Press.
O'Reilly, R.C. & Munakata, Y. (2000). *Computational explorations in cognitive neuroscience: Understanding the mind by simulating the brain*. Cambridge, MA: MIT Press.

6

MACHINE PHYSICALISM

Introduction

Physicalism as a philosophical notion refers to the properties of the physical world and possible worlds. The properties of the physical world as a physical system are recognized in the science of physics as systems from the properties of the physical world. The physical properties of the physical system are emergent phenomena governed by physical laws and principles.

Physicalism is comprised of distinct possible worlds with a range of mechanisms and their parts. The possible worlds of physicalism are depicted in a range of aggregate parts based on the set relation of physical properties in the possible world. Minimal physicalism is the notion that there exists a set relation of the mind to the physical properties of the world. In minimal physicalism, the physical properties of a system consist of matter bits in the world of space–time. In substance physicalism, the physical properties of a system consist of matter bits and exist as bits of matter in the world of space–time or as the matter bits and bits of matter in the space–time world. In nonreductive physicalism, the physical properties of a system are not reducible. The psychological properties of the system of physical properties are not reducible in matter bits or to the world of space–time or the space–time world. In nonreductive physicalism, the systems from the properties of the physical world are not reducible.

In reductive physicalism, the physical properties of a system consist of psychological properties that are reductive. The physical properties of psychological properties comprise kinds and types of physical properties. In type physicalism, the physical properties of a system of psychological properties refer to kinds or types of physical properties or physical property kinds or types. Every psychological property kind is identical or reducible to every physical property kind. Every psychological property type is identical or reducible to every physical property type. In token physicalism, the physical properties of a system of psychological properties refer to unique physical properties. Each physical property of a psychological property is a distinct physical property of an event or state.

In physicalism, the physical properties of the world are parts and particulars of mental phenomena. Mental phenomena consist of sensations that have characters that feel or look or appear in a qualitative way. The qualities of sensations are the experience of being or the phenomena of the state of being. The qualitative way of sensations with characters is a range of phenomena. The feel quality is a phenomenal state of experience as a qualitative state of being. The look quality as a state of being is a phenomenal experience. The appear quality is a state of being that is that range of phenomena.

Physicalism is the postulation that all things in the world have a physical nature and that all things in possible worlds are physical. The world as a physical system has physical properties. All things in the physical world are recognized as physical entities. Physicalism as a description of possible worlds shows a range of mechanisms and their parts. Physicalism as a description of a possible world consists of the physical properties of the mind.

Machine physicalism is the expansion of machine functionalism into the physical system of computation in possible worlds. Machine physicalism entails the functional equivalence of the computation of minds and machines as a demonstration of the physical implementation of functional role in biological and computing machines. Machine physicalism broadens the notion of machine functionalism into physicalism to detail the causal power of the parts and properties of the physical system for the intentionality of intelligent design in possible worlds.

Machine Physicalism

In machine physicalism, the physical properties of the world comprise the phenomena of computation. Mental computation consists of the range of programs for symbolic production. The computing machine as a system of physical properties and particulars for information production sets a standard of criteria for performance. Machine computation as a physical system describes a space–time world as a model net of programs for information production. In the space–time world, machine computation as information production is the production of a set of distinct physical property types. Mental computation is identical to every physical property kind. Machine computation as matter bits is identical to every physical property type in the space–time world.

Machine physicalism as a postulation of the physical system and all things in the possible worlds that are physical entails the intentionality of intelligent design in possible worlds. Machine physicalism suggests that the physical parts and properties of the physical system are autonomous and self-organizing. Machine physicalism comprises the physical parts and properties of computation that are emergent phenomena and performance production as fundamental for possible worlds.

The mind as comprised of mental phenomena consists of mental and physical properties. The mind consists of kinds of psychological property and physical property. The psychological property as a kind of physical properties in the world of space–time or the space–time world. Mental properties and physical properties as a set relation of psychoneural identities are a physical instantiation of the change in spatiotemporal properties that comprises the mind.

Machine physicalism as a philosophical stance posits that the mind is a computing machine with multiple realizers in a possible world. Machine physicalism is a notion of the computing mind as a physical system into possible physical worlds. The notion of the mind as a computing machine with a functional role describes the emphasis placed on the computation of the mind based on its physical realization. Machine physicalism assumes that the computing mind as a physical realization in the physical world is a possible world.

Machine physicalism is a description of a possible world as an organized system with a range of mechanisms and causal interactions. Machine physicalism refers to the physical property of the mind in the physical world as a possible world that acts as an organized system. The physical property of the mind as a computing machine consists of physical particulars and properties. For every particular or property in the actual world, there exists a particular or property in the physical world.

Machine physicalism is an approach to the determination of how everything in the world is physical. In machine physicalism, the physical property of the mind as a computing machine in a physical world is a possible world. All of the physical property of the mind and its computational mechanisms correspond to the mental property of the mind. All of the physical property of the mind in a physical world and its computational mechanisms correspond to the mental property of the mind in a physical world.

Computation and Machine Physicalism

The computational components of machine physicalism consist of the computational principles of the mind in a physical world. The computational components of machine physicalism place emphasis on the computational principles of mechanisms and their causal relations in mind as a physical world. Machine physicalism expands the broader notion of physicalism to include that all of the physical properties of machine computation in the world are physical in nature.

Reductive physicalism refers to the physicalism in set relations of physical properties in the world. The physical properties of set relations in the world comprise in the range of aggregate parts of possible worlds. The range of aggregate parts of physical properties in possible worlds comprise the matter bits and bits of matter of a physical system in the world of space–time and in the space–time world.

Multiple constraint satisfaction refers to the satisfaction of constraints of systems that demonstrate graded, parallel performance. The computational principle of multiple constraint satisfaction is the problem solving through response or interpretation in an activation state that is overall level of satisfaction of many constraints from environmental inputs and learning nets. Multiple constraint satisfaction is the computational principle in mental computation that illustrates the activation state of harmony or goodness as a physical state of property in the physical system. The activation state as a physical state of property in the physical system is the physical part of the physical property that is settled as a matter bit. The activation state as a settlement of matter bit is an optimal state of harmony or goodness in the physical property in the self-organizing system.

Local attractor dynamics are the states of stable activation of networks to a local and global minimum in the physical system. The attractor states maximize the harmony of the network to the optimal state of harmony. The attractor states as a physical state of activation in the physical system illustrate the transition of physical states to a harmony or goodness state of local or global minimum in a physical system. The attractor states as stable activation states optimize to a state of harmony or goodness in the physical property of the self-organizing system. Local attractor dynamics as part of the computational principle of multiple constraint satisfaction illustrate that the network activation of attractor states of internal and external to an optimal state comprises the part properties and the physical property of possible worlds.

Large-scale distributed representation is a component of a computational principle of knowledge representation. Large-scale distributed representation is the distribution of knowledge representation in physical implementation. Large-scale distributed representation refers to the distribution of knowledge across different large-scale brain areas. Multiple brain areas contribute to the large-scale distribution of knowledge representation as units of knowledge. Large-scale representations of knowledge refer to the large-scale properties of brain areas. Large-scale distribution of knowledge representation as a computational component illustrates that the physical property of the physical system is comprised of physical part arrangements or patterns of physical parts. The arrangement of physical parts as large-scale distributed representation comprises the matter bits of mechanisms of the biophysical system. The large-scale distributed representation of knowledge acts as mechanisms of the biophysical system in possible worlds.

Functional specialization is the specialization of mental functions in the organization and structure of the nervous system. Functional specialization refers to the automatic encapsulation of the mental function in focalized brain regions of the nervous system (Fodor, 1983). Functional specialization suggests domain specificity of mental functions within informational processing of neural mechanisms in cortical layers. Functional specialization as a computational principle suggests a functional role of mental function in neural mechanisms that is automatic. Functional specialization consists of the set relations of mental functions to neural mechanisms in specific focal brain regions of the nervous system.

Functional specialization demonstrates the biological plausibility of aggregate parts in the information-processing mechanisms of the biological machine. Functional specialization suggests that the mental computation of biophysical mechanisms in the biological machine is a part of reductive physicalism. Mental computation of the biological machine shares the part properties and property parts of functional equivalence to the computing machine. The functional specialization of mental computation implies an equivalence in standard of task function and performance that is fundamental to minds and machines. Functional specialization illustrates the importance of performance in the physical implementation of mental computation.

The physical implementation of mental computation entails the physical properties and part properties that comprise the nervous system as a biological machine in the physical system of possible worlds. The intentionality of design in the physical implementation of mental computation is the part properties of minds and property parts of machines in possible worlds. The emergent phenomena of part properties and the intelligent design of physical properties for performance illustrate the fundamental units of matter bits in the world.

Machine physicalism implies the functional equivalence of minds and machines exists as an intentionality of the design in the physical implementation of computation. Mental computation as emergent phenomena is part of the physical system of the space–time world. Machine computation as the performance of the physical properties of the physical system shows the matter and physical state transitions of the physical properties of the physical system.

Machine Physicalism in Culture

Machine physicalism refers to the physical properties and physical parts of computation of the physical system. Machine physicalism entails computational performance for information production at the cultural level. Machine physicalism implies that information production of the cultural system is the cultural property of the physical system. Complementarily, machine physicalism suggests that the information production of the cultural system is the physical property of the cultural system level.

Computational principles guide machine computation and information production of the cultural system level. Cultural computation refers to the multilayer strategic capabilities for information production at the cultural system level. Cultural computation is a component of information production for the development and implementation of strategies for cultural security. Cultural security protects the valuation or worth of cultural property. Cultural capital refers to the valuation or worth of cultural property.

Strategic cultural security regulates and protects the development and implementation of initiatives and programs at the cultural system level. Cultural programs include the development and implementation of programs for the empowerment of individuals and groups to contribute to the cultural production of resources and capabilities of the state and its citizens. Cultural participation in the programs of cultural communities ensures the identity development of individuals in their ethnocultural groups. Cultural participation strengthens the relations of ethnocultural groups.

Strategic cultural security protects the cultural capital of cultural property. Strategies of cultural security include the design of prevention and intervention strategies for the protection of cultural capital and cultural development. Prevention and intervention strategies that protect cultural capital seek resources that ensure protection from threats to human security. The design of prevention and intervention strategies for the protection of cultural property ensures the environmental conditions that protect human development.

Machine physicalism produces cultural computation based on computational principles and the physical laws of the natural world. Cultural computation based on computational principles is the simulation and construction of programs for information production for the cultural system level from machine computation. Cultural computation from machine computation is the computation of biological machines and digital computers at the level of the cultural system. The cultural computation of biological machines is the emergent phenomena of cultural property and its parts an organized system of the organism. The cultural computation of biological machines as emergent phenomena is a level of complexity in the self-organization of living systems.

Cultural property consists of the physical property of cultural and physical systems. Cultural property is related to the physical property of culture as an organized system of aggregate parts. The cultural property of the physical system entails the information production from mental and machine computation. The cultural property of the physical system consists of the intelligent design in physical properties and physical particulars. The physical properties and physical particulars of intelligent design are the aggregate parts for the construction of cultural property in the physical system. The cultural property of the physical system in intelligent design performs in the space–time world. The cultural property of intelligent design in the physical system sets standards of performance in the space–time world.

Cultural property and mental property as physical property comprise the cultural computation of the physical system. Cultural computation of the physical system is the set of physical properties and physical particulars that are the aggregate parts of mental computation at the cultural level. Cultural computation of the physical system consists of the cultural net and cultural programs for information production.

Cultural computation as physical property describes the computation of the cultural system in the science of physics. The computation of the cultural system in the science of physics is the set of physical properties and physical particulars as every property kind of the cultural system. The computation of the cultural system in the physical system is the set of physical properties and physical particulars in the cultural system. Cultural computation as physical properties and physical particulars comprises the possible worlds of the cultural system. The possible worlds of the cultural system produce cultural computation in the information production of the physical system.

Cultural computation as physical mechanisms and their computational principles refers to the range of mechanisms and their parts in the physical system of possible worlds. The cultural system in the science of physics is the description of the physical laws of the world and its computational principles at the cultural level. Cultural computation in the physical system is a description of mechanisms and principles of the physical system. The computational principles of cultural computation are the description of the organized system and its parts in the world. The cultural computation in the physical system of a possible world consists of the intelligent design of physical properties and physical particulars of the physical system.

Cultural computation in computing machines is the system-wide performance of programs for information production in the cultural sphere. Cultural computation consists of the computational models and machine learning algorithms for the simulation and construction of cultural production. Cultural computation of computing machines performs the simulation of the traditions, rituals, and values of the cultural system. The automation and control of information production at the level of the cultural system ensure the protection of cultural property. The regulation of internal and external information production at the level of the cultural system defines the standards of production and the development of programs in the cultural sphere.

The cultural computation of machines describes the capabilities for system-wide performance and the protection of information production at the cultural system level. Cultural programs provide the resources that protect the cultural property of minds and machines. The development of cultural programs ensures the performance of cultural programs for cultural participation and cultural life in the cultural sphere. The protection of cultural property entails programs that implement policies and procedures to promote cultural diversity, equity, and inclusion in the public sphere. The protection of cultural property is the control of societal perception and the production of cultural patterns of thought in the world. The development and implementation of cultural programs build the system-wide capability of individuals, groups, and nations for strategic cultural security.

Culture and Machine Physicalism

The cultural level is comprised of a range of phenomena that can be described as a physical system. The range of cultural phenomena consists of the generation of cultural patterns of thought in the mind and the machine production of information at the cultural system level. Cultural dimensions and processes depict levels of the cultural system in internal and external information production. Cultural systems control and protect information production for purposes of strategic cultural security.

Cultural phenomena can be described as physical properties and particulars of the physical system. The set relations of cultural property as physical property comprise the parts of the organized system. The cultural phenomena as emergent phenomena in the natural world relate to mental computation and its functional role in the world. Cultural variation in thought demonstrates the importance of functionalism in behavioral expression.

Cultural patterns of thought and reason suggest that physical laws and principles govern mental phenomena at the cultural level. Cultural mental computation describes the computational principles of thought and reason at the cultural level that guide societal perception and cultural patterns of the mind. Cultural mental computation consists of a range of mental phenomena from cultural perception and learning to feeling and conscious intentionality. Cultural patterns in thought and reason emerge from the influence of culture on the information processing of neurobiological mechanisms and related parts. Cultural patterns of thought and reason demonstrate the biological plausibility of machine computation at the level of the cultural system.

Culture in machine computation consists of the physical parts and particulars that are an autonomous and self-organizing system. The cultural system as autonomous and self-organizing describes the functional performance of mental and machine computation. The cultural dynamics in computation refer to the network activity and flow of information in matter and energy. The cultural dynamics in mental computation consist of the activity of neural networks and the information flow across levels of processing in biophysical mechanisms of the nervous system. The cultural dynamics in machine computation are the cultural patterns of activity of artificial neural networks and the information flow in the matter and energy of cultural artificial life. The cultural dynamics of machine computation in cultural artificial neural networks are networks of activation states to states of harmony or goodness in the complex living system. The cultural dynamics in artificial intelligence regulate the automated production of cultural information.

Cultural brain dynamics entail the physical parts and particulars that comprise the autonomous self-organization of natural and artificial living systems. The functional equivalence of mental and machine computation implies that the cultural system consists of physical parts and particulars that have a functional role. Mental computation to machine computation in physical parts and particulars are elemental units as a flow of information in matter and energy. Machine computation to mental computation in physical parts and particulars are matter bits as a flow of information. Cultural states as mental computation entails a local minimum in the brain dynamics of optimal states. Cultural states as mental computation to machine computation comprise a range of local attractor dynamics as an optimal state. Cultural states as machine computation to mental computation entail a local minimum in the local attractor dynamics as an optimal state. Cultural states of physical parts and particulars perform functional roles for the simulation and construction of optimal states of harmony or goodness as cultural patterns of thought.

Cultural dynamics as brain dynamics refer to the activity of cultural neural networks. Cultural neural networks describe the patterns of activation states to an optimal state that defines a boundary in the information flow of the cultural system. The optimal state in the cultural neural network is the activation state that satisfies the multiple constraints of the cultural system. Environmental inputs affect the local attractor dynamics towards the optimal state of activation that is a state of harmony or goodness in the cultural system. Understanding cultural dynamics as brain dynamics provides insight into the generation of the state of harmony in cultural neural networks. Computational modeling of cultural neural networks contributes to the simulation and construction of the states of harmony or goodness as information production and in cultural patterns of thought.

The cultural dynamics of living systems as autonomous self-organization are comprised of emergent phenomena and complex systems. Cultural dynamics of machine computation play a functional role as environmental input into activation states of cultural nets. Cultural states of machine computation as environmental input consist of the activation states of cultural nets as information flow in matter and energy. Cultural states of machine computation as environmental input are the information production of complex systems. Cultural dynamics of living systems as an autonomous self-organized system perform simulation and construction for information production in the physical system.

Machine Physicalism and Cultural Computation

Early notions of the physical system of computation at the cultural system level is comprised of computing machines that perform symbolic production. The performance of computing machines demonstrates in detail a simulation of the formal procedures of mental function as machine output. The performance of computing machines for machine output introduced the notions of the functional equivalence of minds and machines. The functional equivalence of minds and machines implies that the physical system of computation in living systems is complex, autonomous, and self-organizing. The universal programming of computing machines demonstrates the range of performance across levels of living systems.

The physical system of computation in living systems performs information production across system levels. The computational components of machines perform the design and automation of information production across levels of living systems. The simulation and construction of information production across levels describe the dynamics of living systems in terms of flow of information as matter and energy. The physical system of computation as physical properties and particulars is autonomous and self-organizing. The autonomous dynamics of living systems generate patterns of information flow as cycles in living systems.

The physical system of computation entails the range of performance at the cultural system level of living systems. The computational components of machines perform the design and automation of information production at the level of the cultural system. The simulation and construction of information production at the level of the cultural system consist of the cultural dynamics of living systems in terms of flow of information as matter and energy. The physical system of computation as physical properties and particulars at the cultural system level is autonomous and self-organizing. The autonomous dynamics of living systems generate cultural patterns of thought as information flow as part of the cycles of living systems.

The physical system of computation as physical properties and particulars at the cultural system level consists of the programs for cultural development. The autonomous cultural dynamics of living systems perform the design and automation of programs for cultural development. The implementation of programs for cultural development consists of the information production of resources for system-wide performance at the cultural level of living systems.

Programs for cultural development provide resources for the coordination of participation in cultural life. The coordination of participation in cultural life consists of the cultural activities that maintain tradition in the cultural sphere. The maintenance of cultural heritage and cultural identity contributes to the participation in cultural life and the maintenance of tradition in the cultural sphere. Programs of cultural development contribute to societal perception in public life and the cultural patterns of thought in the public sphere. Programs of cultural development contribute to cultural advancement in the public sphere.

The system-wide performance at the cultural level of living systems reflects the coordination of information production of autonomous, self-organizing states. The states of the physical system as a living system produce autonomous dynamics as cultural patterns of thought. Cultural patterns of thought guide societal perception in public life. Autonomous dynamics in the cultural sphere produce the cultural symbols, heroes, rituals, values, and practices for the maintenance of tradition and participation in cultural life.

Autonomous dynamics in the cultural sphere produce cultural patterns of thought that coordinate participation in cultural life and the maintenance of tradition. The autonomous dynamics of the cultural level of living systems are complex and self-organizing. The autonomous dynamics of cultural patterns of thought are self-organizing states that are coordinated information production.

The functional role of computation at the cultural level is the performance of information production. The computing machine at the cultural level demonstrates functions for the design of programs for information production. Cultural computation regulates information production in the cultural sphere. Cultural computation is part of the intelligent design of autonomous dynamics in the cultural sphere. Cultural computation as the cultural property and cultural capital in the computing machine is the design of programs for human security.

Cultural machine computation is the functional programming of the cultural computing machine. The programming of the cultural computing machine shows the functional equivalence of minds and machines. Cultural mental computation is comprised of the functions of mentality for the generation of cultural patterns of thought. Cultural mental computation as emergent phenomena is part of the complexity of living systems. Cultural mental computation consists of the cultural property as cultural capital in human development.

Cultural neurocomputation is the set of relations of mental and neural property of computation at the cultural level. Cultural neurocomputation relates to the functional properties of the structure and organization of the nervous system of the cultural level. Computational modeling of cultural neural networks describes the patterns of activation states from the bidirectional functional connectivity of brain regions. Cultural neural networks generate patterns of autonomous brain dynamics at the cultural system level. Cultural influences on neural information-processing mechanisms identify patterns of activation in functional networks of brain regions.

Cultural patterns of thought are instantiated in cultural neural networks as distributed representations. Distributed representations of informational content in cortical layers and topographical maps illustrate the patterns of activation states in the functional architecture of brain areas. The activation patterns of cultural neural networks entail the physiological properties of systems and networks. Cultural patterns in neural information-processing mechanisms affect the rates of neurotransmission in information-processing mechanisms at the molecular and cellular level. Cultural influences in neural networks modulate the activation states in pathways of neurotransmission.

Computational modeling of cultural artificial neural networks consists of the simulation and construction of the patterns of activation states for the functionality of cultural nets. Artificial neural networks describe the cultural patterns of thought of the cultural neural network as a cultural model net of activation states. Cultural artificial neural networks simulate the design of cultural patterns as information representation. The simulation of cultural artificial neural networks contributes to the construction of cultural artificial neural structures. The construction of cultural artificial neural structures comprises the design of computational components for cultural neurotechnology.

Cultural neurotechnology describes the application of cultural neuro-computation for the design and construction of cultural devices and cultural computers. The design and construction of cultural devices and cultural computers that respond in simulation and in the real world demonstrate the function of circuits and neurons as interacting and intrinsic properties of complex and emergent living systems. The simulation and construction of cultural devices and cultural computers demonstrate the performance capabilities of cultural computation.

The simulation in cultural devices and cultural computers contributes to the construction of cultural life in living systems. The simulation in cultural devices and cultural computers is part of the construction of a cultural model net and net-to-net interaction in possible worlds. The interfacing of cultural devices and cultural computers through the simulation of the real world contributes to the construction of real-world properties in the world of space–time and the space–time world as in real-world interactions.

The net-to-net interactions are component parts of the simulation and construction of artificial life systems in possible worlds. The interactions of net to net construct particular properties of matter bits. Net-to-net interactions and their particular properties comprise the matter bits of complex artificial life systems. The interactions of complex artificial life systems are autonomous and self-organizing. The artificial life system as a living system constructs cycles of patterns of information flow in physical particulars and properties of the physical system.

Cultural computation as machine property is the information production from machine computation. Cultural computation as machine property refers to the functional performance of computing machines in the cultural system. Machine property as part of cultural property is the performance of computing machines at the cultural system level. Machine capital as a part of cultural capital is the valuation or worth of information production from machine production. The protection of machine capital in computational components of a complex system is the purpose of machine security. Machine security as part of cultural security contributes to the strategic capabilities of cultural computation and the performance of programs of human security.

Machine physicalism postulates the design of the physical system for cultural computation in possible worlds. Machine physicalism maintains the parts and properties of the physical system for computation at the cultural system level. The cultural computation of possible worlds performs as the intention of intelligent design.

Conclusion

The physical world is comprised of physical parts and particulars that are fundamental components of mental life in the cultural sphere. Mental life as natural phenomena is related to emergent mechanisms and the physical parts and particulars of possible worlds. Mental life in the cultural sphere reflects the intentionality of intelligent design in the world. Cultural life as cultural patterns of thought and performance of cultural production promotes cultural diversity, equity, and inclusion in the public sphere. The cultural level of living systems reflects the importance of cultural and biological variation in mental life.

Cultural computation in machine physicalism as the performance of information production illustrates the functional role of physical parts and particulars at the cultural system level. Machine physicalism as information production from computation demonstrates a functional role of minds and machines at the level of the cultural system. Machine physicalism underscores the functional role of mental computation in possible worlds and the importance of intentionality in the intelligent design of cultural property in the physical system.

The functional equivalence of minds and machines in computational performance underscores the importance of intentionality in the design of cultural computation. The design of computers that perform cultural programs contributes to the resources for empowerment and cultural development. Cultural programs demonstrate the universal purpose of programming computing machines. Cultural computers illustrate the functional role of machines for the performance and production of mental life in the cultural sphere. Cultural life encompasses the cultural activities and traditions that broaden a sense of cultural participation in the life of the mind.

Reference

Fodor, J. (1983). *Modularity of mind.* Cambridge, MA: MIT Press.

Further reading

Kim, J. (2011). *Philosophy of mind (3rd edition).* Boulder, CO: Westview Press.

7

COMPUTATIONAL THEORY OF MIND

Introduction

The philosophical inquiry into the mind broadens into computation with the contemporary metaphor of the mind as a computing machine. Philosophical postulation into the nature of the mind as computation has cultivated several theoretical stances regarding the relation of the mind to the brain, the function of the mind as a computer, and the actuality of the mind as a form of computation. Computational theory of mind represents a conceptual shift in philosophical inquiry from logical positivism to empiricism in pursuit of a complete and satisfactory explanation of the fundamental nature of the mind.

Computational Theory of Mind

Computational theory of mind consists of the mental and physical property that produces the functional output of mental state understanding that is deterministic. Computational theory of mind includes the mental and physical events that demonstrate the processing of information to produce a predetermined output or goal state from the mind. The mental and physical events of a predetermined output or goal state are consistent with the concept of rule. The implementation of rule through a deterministic computational theory of mind aims to produce mental and physical property that assures a given functional output. Computational theory of mind assures a social inference of the mental state of another mind based on interaction with the social environment.

At the computational level, mental and physical states that are deterministic are comprised of functional property consistent with a particular functional task and its components (Marr, 1982). The functional property produced from a particular task and its components may be predetermined. A functional task of mental state understanding may consist of subsets of mental and physical states for computation of social inference organized into parts of its subcomponents. For instance, the functional task of mental state inference from social visual cues consists of subsets of mental and physical states that are subcomponents, including the representation of sense data from a face, the transformation of sense data from facial input into configural and featural processing, and the encoding of representational face data into memory. The functional task of mental state inference unfolds in a predeterministic fashion such that a subcomponent given a particular sensory input or its transformation produces an internal representation of a particular social inferential output. Mental state understanding can be considered deterministic at the computational level based on functional tasks and their components.

Mental state understanding may be deterministic at the level of functional task to a specific cultural system. The functional property that is produced from a particular task and its components may be predetermined cultural property. The cultural property of mental state understanding consists of the cultural functional property, for instance the functional task of mental state inference and its components of social decoding, and the cultural sense data. Across cultures, functional property and its specific task components are distinct sets of physical states and internal states of representation. For instance, cultural variation in mental state inference consists of distinct subsets of mental and physical states and their transformation functions. Thus, mental state understanding at the computational level is a functional property of cultural systems.

Mental and physical processes that are deterministic include algorithms that are defined through functional property are consistent of a particular input and output. The generation of functional property of a given output with a particular input may require transformation of the particular input into the functional output. The generation of functional property for mental state understanding includes the transformation of sense data into the representational content of internal mental states. For instance, the perception of cultural patterns within the eye region of the face inputs cultural sense data that transforms into internal representations of culture-specific mental states.

The algorithms of mental state understanding may be deterministic to a specific cultural system. First, cultural property consists of tasks for mental state understanding that may be defined through constituent parts as culture-specific functional property. Second, the generation of the functional property of a given output with a particular input may be performed as an algorithm with varying ease or to a greater extent within a specific cultural system. Cultural variation in mental state understanding at the levels of computation and algorithm may perform independently. Cultural variation in mental state understanding observed through the ease or availability of culture-specific algorithms illustrates the determinism of cultural systems.

The production of functional property of a given output with a given input may require the implementation of a rule such that a particular input produces a functional output. The production of functional property that is consistent with theory of mind includes the automatic and controlled processes of social reasoning. The production of mental state understanding of a given output with a particular input from a culture-specific task or algorithm may consist of social thought that is controlled and deterministic of cultural systems. Culture-specific algorithms for theory of mind ensure the production of controlled social thought as an implementation of rule within a cultural system.

Mental and physical events that comprise a computational theory of mind include a physical implementation. The physical implementation of a computational theory of mind consists of patterns of mental and physical events that are located within specialized functional brain circuitry for mental state understanding. Specialized brain circuitry for theory of mind includes the set of brain regions that are necessary and sufficient to produce the internal representations of the mental state of others. The functional specialization of brain circuitry for theory of mind demonstrates the biological plausibility of computational models of theory of mind.

Computational modeling of theory of mind may consist of sets of tasks and constituent parts for inferring the mental states of others. Computational models may characterize the functional tasks of social inference as comprised of several components, including the internal representation of a social identity or another mind with different types of social mental states, including emotion, social attribution, and belief, among others. The content of the functional output of social inferential processes consists of the physical and representational states of mental state inference. Cultural systems of social thought comprise the set of physical and representational states of social inference that observe culture-specificity across levels of analysis.

Cultural systems of social thought may produce functional output that is culture-specific. Cultural variation in mental state inference, such as emotion recognition and theory of mind, consists of the sets of physical and internal representational states that show distinct patterns of functional output based on culture-specific information processing. Cultural variation in levels of mental state understanding reflects a systematic mapping of the states of the cultural system to the internal representations of another mind. Cultural variation in mental state understanding reflects culture-specific valuation in functional property.

The physical implementation of computational models of theory of mind is comprised of neural systems of social cognition. Brain regions for automatic processes of social perception include perceptual areas of the primary visual cortex. Cortical regions in primary visual cortex, including the fusiform gyrus and extrastriate visual areas, are specialized for the automatic and rapid processing of social information from visual cues. The fusiform gyrus consists of subregions that are highly responsive to the perceptual processing of faces. The extrastriate areas are comprised of subregions that are specialized for the perceptual processing of bodies. The perceptual processing of social visual cues, such as faces and bodies, within the primary visual cortex occurs automatically and rapidly within milliseconds of input from sense data.

Brain regions for automatic processes of mental state understanding consist of subregions of the superior temporal sulcus. In the Reading the Eyes in the Mind task, perceptual processing of social visual cues from the eye region generates social inferences of the mental states of others (Baron-Cohen, Wheelwright, Hill, Raste, Plumb, 2001). Task performance in the Reading the Eyes in the Mind task demonstrates the information-processing steps of social inference for understanding the mental states of others.

Brain regions for controlled processes of social reasoning include the prefrontal cortex and interconnected subregions. The prefrontal cortex is a brain region necessary for the regulation of automatic processes into social reasoning. The prefrontal cortex is a top-down neural mechanism to regulate the functional activity of brain regions interconnected to automatic processes of the mind and underlying neural mechanisms. The regulation of functional activity of brain regions reflects a set of mental and physical events that are consistent with patterns of conscious social thought.

Cultural variation in neural mechanisms of conscious social thought illustrates a physical implementation of cultural systems. Culture-specific patterns of conscious social thought within specialized brain circuitry reflect the cultural property generated in an emergent fashion from the biological organism. Cultural-specific patterns of neural activation and its relations with conscious social thought are physical and mental events that show conservation of the cultural system in the biological organism.

What are the Principles of Computational Theory of Mind?

Principles of computational theory of mind are consistent with those of computational social cognition. Computational principles of cognitive processes for understanding other minds, or social cognition, consist of the tasks and constituent parts for knowledge and understanding of the mental states of self and others. The general properties of computational social cognition consist of information-processing mechanisms for social inference that are comprised of structural and dynamical aspects (O'Reilly & Munakata, 2000). General principles of computational social cognition observe structural and dynamical aspects of information-processing mechanisms.

Structural principles of computational social cognition consist of hierarchical information processing. Hierarchical information processing refers to layers of specialized information representation across subordinate and superordinate levels. Hierarchical information processing in computational social cognition is comprised of layers of representation for the understanding of mental states and knowledge representation of self and others. Understanding mental states refers to the hierarchical layers of representational content of other minds, from perception and emotion to attribution and belief. Theory of mind consists of the representational content of mental state inference that is emergent or constructed from the social information processing of the self and others through interaction with natural or man-made social environments, respectively.

Computational social cognition relies on functionally specialized pathways for the information processing of mental states. Functional specialization of pathways for theory of mind is neural mechanisms dedicated to the understanding of mental states, including false belief. Functionally specialized pathways for theory of mind are interconnected brain regions, distributed across different processing pathways (Saxe, Carey, Kanwisher, 2004). Subordinate processes of theory of mind, such as false belief understanding, may demonstrate modular properties within specialized neural processing pathways.

Layers of information processing for theory of mind consist of a hierarchical structure. Hierarchical layers of information processing perform the transformation of social sense data as input into the production of motor output or interpretation of sensory input consistent with social environmental input. Representations of social mental states consist of the transformation of sense data from social input (e.g., human mind) into the interpretation of the social input (e.g., mental state) or the production of motor output as social behavior (e.g., expression of the eye region). The transformations of social sense data observe spatial invariance and are performed based on the dimensions or aspects of the social input across all levels of processing. Cultural systems contribute to the information transformation of mental representations of social sense data through the amplification of features or properties of internal representations that are meaningful as social input.

Specialized pathways or streams of processing are important for social information processing. Social information processing from representations of social percepts are sequential and require the transformation of social sense data into representations of information across multiple properties. The transformation of social sense data into the property of social categorical representation relies on a layer of processing. The representation of social categorical information consists of the encoding of specific properties that aggregate the lower-level social detail into an overall social visual information-processing stream.

The principle of multiple constraint satisfaction describes the production of bidirectionally connected neural networks. Social neural networks will produce an activation state from social input data that satisfies constraints from environmental inputs and learned weighting. The social neural network will produce a rapid pattern of an optimal state consistent with or based on familiar input. Iterative searching of the social neural network may similarly produce an optimal state based on novel input. Due to attractor dynamics, the social neural network will produce a completed pattern as a stable activation state based on a range of initial social input.

Specialized pathways for theory of mind may demonstrate multiple constraint satisfaction. The information processing of mental states is informed across multiple pathways to perform higher-level interpretation and produce meaningful output that is mutually constrained and satisfies multiple conditions. Higher-level processing pathways contribute to the processing of functions that change the interpretation of social input or control the production of motor response comprising social behavior.

Dynamic principles of computational social cognition refer to the bidirectional connectivity of networks. Social information processing consists of bidirectional network connectivity under multiple constraint satisfaction. First, simple social percepts or social sense data may produce a response or motor output through feedforward processing, demonstrating an optimal state without feedback. Second, complex social percepts or social sense data may require bidirectional feedforward processing relying on feedforward and feedback inhibition with interactive searching to perform pattern completion that satisfies constraints. Feedforward nets that produce multiple activation states consist of interpretation of the sense data, while feedback inhibition that performs activity patterns leads to maximal satisfaction of constraints before response selection. Feedback inhibition may consist of patterns of activity that are consistent with top-down or iterative processing on social percepts or social sense data. Feedback inhibition may be useful for maximization of the overall harmony state of the network.

Dynamical principles of computational social cognition characterize the goal of patterns of activity of networks and their optimal state. The patterns of network activation perform computation to satisfy constraints from environmental input and the weights and activation states of the nets towards a harmony state or level of satisfaction. The goal of the network activity states is to harmonize the activation state towards an overall lower level of energy. A state of the network in satisfaction has a lower energy state. Less interaction of network states reflects a lower energy state of the network itself. The most harmonious states of network activation refer to attractor states, or states of network activation that define a local minima of the energy function.

Computational models of social cognition may characterize patterns of network activation and their overall level of satisfaction or harmony state. Patterns of activation in neural networks of social cognition characterize neural activation within subregions of prefrontal, temporal, and parietal cortices and their interconnectivity. Social neural network activation reflects the transformation of social environmental input (e.g., eye region input) into internal representations of mental states (e.g., mental state attribution). The transformation of social input to output comprises transitions of the social neural network to lower states of energy. The production of the internal representation of mental state attribution reflects a functional state of the social neural network and its activation in a state of harmony.

Culture and Computational Theory of Mind

Culture refers to the shared meaning systems of groups of people defined by ancestry, language, customs, geographic origin, and ethnic heritage. Cultural dimensions consist of systems of values, practices, and beliefs that guide mental processes and their underlying mechanisms in response to adaptation (Gelfand et al., 2011; Oyserman, Coon, Kemmelmeier, 2002). Cultural orientations reflect the sets of mental processes that define the notion of the self and its relation to the social and physical world (Markus & Kitayama, 1991). Cultural systems of thought guide the metaphysics and epistemology of knowledge of the social and physical world (Nisbett, Peng, Choi, Norenzayan, 2001). Cultural processes reflect the hierarchical structure of systems of knowledge and their interaction with the environmental and individual level.

Dynamic models of cultural processes characterize the construction of cultural constructs from environmental input. Cultural frames refer to the cognitive processes or mindsets that become highly accessible through environmental exposure and task switching (Hong, Morris, Chiu, Benet-Martinez, 2000). Cultural frames reflect the internal representations of distinct systems of domain-specific knowledge that become more accessible with exposure and task frame switching. The patterns of activation of cultural constructs are greater when the salience of domain-specific knowledge is heightened through cultural knowledge or the use of language as a cultural prime.

Cultural identity reflects the formation of self identity through societal practices of belonging and commitment that strengthen the maintenance of one's ethnic heritage. Cultural identity consists of the values, practices, and beliefs that demonstrate the inclusion of the self with others within the ethnic heritage. Belonging to the social group of one's ethnic heritage contributes to a sense of group membership and affirmation of the self within the cultural community (Phinney, 1992). The sharing of social roles with cultural group members and the engagement of habitual cultural practices demonstrate the social coordination of resources and roles for self and others within a cultural context.

Cultural processes are fundamental to the generation and production of cultural adaptation. The evolution of cultural processes refers to the acquisition of cultural capacities in response to environmental demand. Cultural variation in social organization and societal practices reflects distinct use of mental constructs that produce psychological adaptation. Cultural variation in social organization contributes to the acquisition of distinct worldviews and systems of thought. Cultural variation in thought reflects levels of use, function, and accessibility of specific mental constructs (Norenzayan & Heine, 2005).

Cultural systems of thought guide metaphysics and epistemology and knowledge of the social and physical world through development and use. Cultural processes are emergent in the organism through maturational periods of development that demonstrate the acquisition of culture capacities in mental constructs.

The acquisition of theory of mind as a social capacity reflects the developmental maturation of the social mind to understand the mental states of others. Understanding the mental states of others during early development constitutes the acquisition of evolutionary precursors to cultural capacities during development.

Cultural transmission refers to the cultural processes of knowledge acquisition through development and social learning (Boyd & Richerson, 2005). Vertical cultural transmission reflects the hierarchical acquisition of knowledge from genetically related individuals within the social group. Horizontal cultural transmission refers to knowledge acquisition through social learning of unrelated individuals with a cultural model. The persistence of social thought from cultural model to learner refers to the effective transmission of cultural information across individuals.

The evolutionary biases of cultural transmission demonstrate the cultural processes that guide patterns of transmission of information through social learning. Cultural evolutionary processes define capacities of social information transmission that produce levels of variation in social learning (Mesoudi, 2009). Cultural acquisition through conformity refers to the use of social information based on popularity. Model-based biases of social transmission reflect the acquisition of social information from cultural models defined by prestige or success. Content biases can heighten the preferential transmission of social information based on its content. Cultural group selection refers to the biases of social information processing that contribute to intragroup and intergroup processes.

Cultural evolutionary processes of social transmission define patterns of cultural variation in social information processing. Cultural evolutionary biases contribute to the cultural variation in patterns of social learning. Cultural variation in social thought may result from differential levels of model-based social learning across cultural dimensions. Content biases may demonstrate heightened transmission of social belief across social learners due to its content. Cultural group selection may further result in cultural variation due to the salience of conformity and other social information-processing biases for cultural and psychological adaptation.

Cultural variation in social thought reflects levels of processing during mental state understanding. Cultural variation in social thought may arise from different levels of use, accessibility, and function in social information processing across cultural contexts. Cultures differ in their emphasis on the importance of social thought for self and others. Cultural variation in social thought illustrates patterns of processing that differentially contribute to goal attainment for self and others.

Cultural dimensions show differential patterns of social thought. The cultural dimension of individualism and collectivism reflects the differential emphasis of the self and its relation to others in the environment as worldviews (Oyserman, Coon, Kemmelmeier, 2002). Individualism places emphasis on the individuals as independent and unique from one another with concern for autonomy and personal goals. Individualism promotes a positive regard for the self and a self-concept defined through dispositional social reasoning or trait attribution. Individualism encourages emotional expression and the centrality of the experience of the self for causality and social reasoning. Collectivism conceptualizes a worldview defined by duties and obligations of the self with others. Collectivism promotes the social group as central to identity and goal attainment. Collectivism emphasizes the maintenance of harmonious relationships and the regulation of emotional expression for the promotion of ingroup harmony.

The cultural computation of social neural networks illustrates the transformation of cultural sense data into internal representations of social information within the nervous system. Empathy, as the perception and experience of the emotional states of others, is related to patterns of heightened neural activity within this network of affective brain regions (Lamm, Batson, Decety, 2007). Social neural networks for understanding the emotional states of others consist of brain regions within the anterior cingulate cortex, anterior insula, and medial prefrontal cortex, among others.

Cultural variation in social neural networks demonstrates distinct patterns of neural activation across cultural dimensions. Other-focusedness as a component of the cultural dimension of individualism and collectivism refers to the differential emphasis on the perspectives and outcomes of others (Oyserman, Coon, Kemmelmeier, 2002). Cultural variation in functional neural activation during empathy is related to other-focusedness (Cheon et al., 2013). Greater other-focusedness is related to a pattern of heightened neural activation within brain regions of the anterior cingulate cortex, anterior insula, and medial prefrontal cortex during empathy. Cultural variation in heightened neural activation in the social neural network demonstrates the levels of processing for understanding the pain of others. Cultural computation in social neural networks contributes to patterns of social thought of cultural adaptation.

Cultural variation in the neural systems of theory of mind demonstrates patterns of activation that are culture-specific. Cortical subregions of the superior temporal sulcus show greater neural response during the theory of mind task. Patterns of heightened neural activation within the superior temporal sulcus show the functional processing of social visual cues and their transformation into internal representations of social mental states (Allison, Puce, McCarthy, 2000). Cultural group members show heightened neural activation within specialized brain regions for theory of mind consistent with task performance (Adams et al., 2010). The influence of cultural group membership on patterns of neural activation within specialized brain regions for theory of mind suggests cultural biases in social information-processing mechanisms. These cultural biases contribute to the acquisition of social information important to cultural and psychological adaptation.

Cultural specificity in specialized neural regions of theory of mind illustrates cultural computation in the nervous system. The cultural computation for theory of mind is comprised of levels of information processing within functionally specialized brain regions. These physiological structures consist of the representational content of cultural states of theory of mind and their transformations from sensory input. The transformations of cultural sense data into representations of cultural content within the nervous system imply physical states and physical state transitions towards lower levels of energy or a state of harmony.

The influences of culture on computational theory of mind are multifold. Culture consists of distinct computational systems for social processes. Distinct patterns of neural activity for theory of mind constitute a causal explanation for cultural dimensions of social reasoning. The identification of physiological structures for cultural computation defines a spatial location for the mapping of neural states to elements of culture. The mapping of sets of neural states to cultural mental states implies machine functionalism that is algorithmic or rule-based.

At the computational level, culture consists of sets of culture-based functional tasks and rule sets. The culture-based rule sets define deterministic procedures of functional operations. Culture in the mind is comprised of physical states and their state transitions as internal representations of thought. Culture-based rule sets define the set of physical state transitions that comprise the physical realization of cultural patterns of thought. Cultural patterns of thought comprise streams of states of consciousness as the emergent property of real-world experience.

Cultural neurocomputation consists of the representational content of culture in the computational levels of the nervous system. The nervous system is the physical system that performs the computations of cultural processes. The physical states of the nervous system and their physical state transitions perform cultural computations. In cultural neurocomputation, the causal explanations of culture are characterized by the patterns of activity of the nervous system. The patterns of neural activity are the causal explanation of culture. The cultural computations of the nervous system reflect emergent properties of a physical system at the computational level of cultural processes.

Cultural neurocomputation implies that the cultural computer is a biological computing machine. The physical states of the nervous system comprise a biological computing machine for culture. The cultural brain as a cultural computer performs mental functions of cultural processes. The cultural brain implies that laws and principles of the natural world govern the patterns and regularities of its cultural computation. The notion of the cultural mind as a cultural brain contributes to the discovery of cultural processes as a pursuit of a biological naturalism.

Computational Theory of Mind in Culture

The generation and maintenance of culture consist of processes for cultural production and cultural acquisition. The mental and computational property of the nervous system produces cultural property through a sequence of processes within the cultural system. Mental and computational property produces culture through the storage and transformation of informational content that is consistent with the cultural system.

The mental and computational property of the biological organism acquires the cultural property of language through experience and learning. The mental and computational property of the biological organism can produce culture through sense data that serves as input of informational content and undergoes transformation to produce output.

The biological organism can simulate the computational property of a computer program. The biological organism can simulate the computational property of a computer program modeling mental property. The computational property of the biological organism is also computational property. The computational property of the biological organism can be stored or generated as identical to itself, but not to the mental and computational property that is not from a biological organism.

The computational and mental property of the biological organism can produce novel cultural property that is unique to the biological organism. The cultural property of the biological organism is cumulative, building on initial mental and computational states of the biological organism in a rapid fashion. The cultural property of a population of biological organisms is similarly cumulative, capable of building on the initial states of mental and computational property that are aggregate of the population.

Conclusion

Computational theory of mind advances the philosophical notion of the mind as a machine. Computational theory of mind expands the mind as machine metaphor to the computer, such that the mind performs computation as the computer. The philosophical stance of computational theory of mind suggests how the mind as a computer performs the functional task of mentality and thought. The physical realization of the mind as a computer postulates of the biophysical mechanisms at the computational level that generate the processes of the mind.

In pursuit of a biological naturalism to understanding the mind, philosophers and scientists alike have made notable progress in the discovery of the mind as a brain. The biological naturalistic perspective has shifted emphasis from logical positivism to empiricism for theory confirmation. The computational perspective to theory of mind has contributed to the biological plausibility of the mind. Philosophical considerations into the nature of the mind demarcate computational theory of mind as a formidable stance from which contemporary understanding of the mind and computation has flourished.

References

Adams, R.B., Jr., Rule, N.O., Franklin, R.G. Jr., Wang, E., Stevenson, M.T., Yoshikawa, S., Nomura, M., Sato, W., Kveraga, K., Ambady, N. (2010). Cross-cultural reading the mind in the eyes: an fMRI investigation. *Journal of Cognitive Neuroscience, 22(1)*, 97–108.

Allison, T., Puce, A., McCarthy, G. (2000). Social perception from visual cues: role of the STS region. *Trends in Cognitive Sciences, 4*, 267–278.

Baron-Cohen, S., Wheelwright, S., Hill, J., Raste, Y., Plumb, I. (2001). The "reading the mind in the eyes" test revised version: a study with normal adults, and adults with Asperger syndrome or high-functioning autism. *Journal of Child Psychology and Psychiatry*, *42*, 241–251.

Boyd, R. & Richerson, P.J. (2005). *The origin and evolution of cultures*. New York: Oxford University Press.

Cheon, B.K., Im, D., Harada, T., Kim, J.S., Mathur, V.A., Scimeca, J.M., Parrish, T.B., Park, H., Chiao, J.Y. (2013). Cultural modulation of the neural correlates of emotional pain perception: the role of other-focusedness. *Neuropsychologia*, *51(7)*, 1177–1186.

Gelfand, M.J., Raver, J.L., Nishii, L., Leslie, L.M., Lun, J., Lim, B.C., Duan, L., Almaliach, A., Ang, S., Amadottir, J., Aycan, Z., Boehnke, K., Boski, P., Cabecinhas, R., Chan, D., Chhokar, J., D'Amato, A., Ferrer, M., Fischlmayr, R., Fülöp, M., Georgas, J., Kashima, E.S., Kashima, Y., Kim, K., Lempereur, A., Marquez, P., Othman, R., Overlaet, B., Panagiotopoulou, P., Peltzer, K., Perez-Florizno, L.R., Ponomarenko, L., Realo, A., Schei, V., Schmitt, M., Smith, P.B., Soomro, N., Szabo, E., Taveesin, N., Toyama, M., Van de Vliert, E., Vohra, N., Ward, C., Yamaguchi, S. (2011). Differences between tight and loose cultures: a 33-nation study. *Science*, *332(6033)*, 1100–1104.

Hong, Y.Y., Morris, M.W., Chiu, C.Y., Benet-Martinez, V. (2000). Multicultural minds: a dynamic constructivist approach to culture and cognition. *American Psychologist*, *55(7)*, 709–720.

Lamm, C., Batson, C.D., Decety, J. (2007). The neural substrate of human empathy: effects of perspective-taking and cognitive appraisal. *Journal of Cognitive Neuroscience*, *19(1)*, 42–58.

Markus, H.R. & Kitayama, S. (1991). Culture and the self: implications for cognition, emotion and motivation. *Psychological Review*, *98(2)*, 224–253.

Marr, D. (1982). *Vision*. New York: Freeman.

Mesoudi, A. (2009). How cultural evolutionary theory can inform social psychology and vice versa. *Psychological Review*, *116(4)*, 929–952.

Nisbett, R.E., Peng, K., Choi, I., Norenzayan, A. (2001). Culture and systems of thought: holistic versus analytic cognition. *Psychological Review*, *108(2)*, 291–310.

Norenzayan, A. & Heine, S.J. (2005). Psychological universals: what are they and how can we know? *Psychological Bulletin*, *131(5)*, 763–784.

O'Reilly, R.C. & Munakata, Y. (2000). *Computational explorations in cognitive neuroscience: Understanding the mind by simulating the brain*. Cambridge, MA: MIT Press.

Oyserman, D., Coon, H.M., Kemmelmeier, M. (2002). Rethinking individualism and collectivism: evaluation of theoretical assumptions and meta-analyses. *Psychological Bulletin*, *128(1)*, 3–72.

Phinney, J. (1992). Ethnic identity in adolescents and adults: review of research. *Psychological Bulletin*, *108(3)*, 499–514.

Saxe, R., Carey, S., Kanwisher, N. (2004). Understanding other minds: linking developmental psychology and functional neuroimaging. *Annual Review of Psychology*, *55*, 87–124.

Further reading

Laland, K.N., Odling-Smee, J., Feldman, M.W. (2000). Niche construction, biological evolution and cultural change. *Behavioral and Brain Sciences*, *23(1)*, 131–146.

8

SIMULATION

Introduction

The generation and maintenance of culture demonstrate the importance of social processes in the production of mental and computational property of the organism and the world. The mental and computational property of the organism and the world for the simulation of mental and physical states of cultural processes comprises the understanding of mental states of social systems. Mental state understanding consists of multilevel mechanisms for the social capacity to generate and share the mental and physical states of cultural processes. Simulation as a form of understanding other minds facilitates the generation and production of culture in the organism and the world through social processes.

Cultural dimensions generate and maintain states of social processes through interaction with ecological conditions. The interaction of ecological conditions and cultural dimensions produces states of social processes that consist of sets of specialized tasks for cultural and psychological adaptation. The social processes comprise specialized functions that ensure the strengthening of the dimensions of culture. The performance of specialized social tasks consists of an accurate output given a specific input. Cultural dimensions ensure the adaptation of social behavior through social transmission.

Cultural transmission consists of the acquisition of social information across minds through the processes of social learning. Cultural transmission generates and constructs mental state understanding through social learning processes. The generation of mental state understanding from fundamental social relations consists of social experience that is beneficial for the organism. The understanding of the mental states of close others reflects an experience of social learning that is satisfactory and necessary as a causal explanation for social knowledge in the social world.

The cultural transmission of social information through other minds constitutes a form of cumulative social learning. The sharing of mental states across other minds consists of social information that is cumulative across minds from model to learner. The cultural model characterized through social processes demonstrates the capability to share social information in the social learning mind. The social learning mind through the demonstration of its shared understanding with the cultural model contributes to the cultural construction of social knowledge in the world.

Understanding the mental states of biological organisms is foundational to the construction of mental state understanding of computer models. The mental state understanding of biological organisms constitutes the generation of the state of satisfaction from social knowledge in the world. The simulation model of biological organisms generates states of social knowledge that satisfy multiple constraints. The state of satisfaction from the social knowledge of biological organisms is social harmony and reflects an adaptive capability of mental state understanding.

The computational modeling of simulation is for the construction of the understanding of other minds in the world. Understanding mental states from a computer model is a construction from the model components of social knowledge in the world. The computer modeling of simulation is a construction of the states of satisfaction from social knowledge in the model net and the world. The modeling of simulation from biological machine to cultural computer builds from the development of technology for the cultural construction of social knowledge in the world.

The construction of simulation from cultural computer to biological organism consists of the understanding of mental states for the performance output of real-world advantages. The simulation model of a cultural computer shares the performance output from the cultural model net for real-world advantages. The cultural computer is comprised of a cultural model net that performs output of the simulation of other minds and the world for real-world advantages.

The cultural model net and its synthetic components consist of the performance output for the simulation of the world. The simulation of the world from synthetic components is constructed from the modeling of mental state understanding of the biological organism into parts of the synthetic world and its model components. The production of a cultural chip that stores the representational content of other minds in its optimal state is an example of the capability of simulation in a synthetic part. The production of a cultural computer that contains a model net of the representational content of other minds in its optimal state is an example of the capability of simulation in a synthetic device. The cultural computer and its components demonstrate the strategic capability of social performance that is advantageous in real time.

The simulation of the mental state understanding of the biological organism in the synthetic world consists of the construction of technology for cultural development. Technology for cultural development contributes to the construction of the cultural computer for social performance that produces real-world advantages.

The cultural construction of social knowledge in the world builds from the benefits of mental state understanding. Mental state understanding promotes social harmony amongst groups and individuals. The state of other minds in social harmony reflects a satisfaction in the processes of social relations. The sharing of mental states across minds contributes to social cooperation in the social world. The social cooperation of individuals and groups in the social world is necessary for societal achievement in the world.

Simulation

Understanding others forms the basis of the structural aspects of society and social relations. Social understanding is foundational to the structure of societal organization and social relations. The organization of societal roles and responsibilities characterizes the fundamental structure of social coordination within social groups. The structure of social relations amongst individuals and groups depicts distinct relational kinds. Kinds of social relations build from the interaction of others that entail a scope of assumptions of the content and character of the minds of others.

Social interaction builds on the understanding of the minds of others. Simulation characterizes the generation and sharing of mental states of other minds for the purpose of mental state understanding. The simulation of mental states of other minds consists of the mental and physical property that is social. The simulation of the minds of others entails the formation of possible worlds of social knowledge. The interaction of minds in the world suggests the generation of social knowledge in possible social worlds.

The generation of social knowledge contains the representational content and characteristics that are attributable to the minds of others. The possibility of social worlds describes the realm of social knowledge that is generated through the imagination of perspective. Through the generation of perspective, the mind imagines the world as it is and as it ought to be. Through social perspective, the mind generates knowledge of the social world as it is and as it seems to the minds of others.

The generation of perspective in the mind depicts a first-person understanding of the social world. The first-person perspective is an understanding of the world that is singular to that person perspective. The perspective of the first-person is generated through experiential understanding of the social world. The experience of the social world consists of the understanding of the self and others. The recognition in the social world of the self and others and the contingent interaction of the mind in the social world with other minds comprise social experience. The generation of experiential knowledge of self and others builds the content of the social world as it is. The imagination of the social world of the self and others constitutes theory generation of the social world.

Third-person perspective consists of the understanding of the minds of others. The third-person perspective refers to the social knowledge in the world that is constant across experiential understanding. Third-person perspective comprises the knowledge generation in the social world that is independent across other minds and acts as a social reference. The perspective of the third person is accessible to the first-person perspective through social inference. The third-person perspective reflects the continuity of the knowledge of other minds across spatio-temporal dimensions.

The subjectivity and objectivity of conscious experience in the social world refer to the representational content of the first- and third-person perspective. The subjectivity of conscious experience reflects the characteristics of the mind that are private and unique. Conscious experience in its subjectivity consists of the emergent properties of interacting parts and their causal effects. The objectivity of conscious experiences is composed of the representational content that refers to the regularities of social world and in the robust patterns of understanding in other minds in the natural world. Conscious experience in its objectivity refers to the regularities in the social world that describe the mechanism of the higher-level features of social experience that arise from a configuration of parts within a system in the social world. Objectivity in conscious experience describes the generalizations in the understanding of other minds that are true in the natural world.

Theory generation in the social world consists of the analysis of the patterns of mechanisms in conscious experience. Theory generation in the social world considers the patterns of mechanism from the subjective and objective parts of conscious experience and their causal effects. Theory generation in the social world consists of the analysis of the patterns of mechanism from the subjective parts of conscious experience as emergent properties of the mind. The generation of knowledge from the subjectivity of conscious experience consists of the patterns of mechanism from the interaction of independent parts. Theory generation in the social world from the objective parts of conscious experience consists of the robust patterns of the mind in the natural world. The generation of knowledge from the objectivity of conscious experience is comprised from the regularities in patterns of nature that hold.

The imagination of subjectivity in conscious experience generates the content and character for patterns of the mind in the natural world. The subjective imagination consists of the regularities in conscious experience that are independent from parts of the social world. The independence of patterns and regularities in conscious experience generates content with the interaction of the mind in the natural world. Conscious experience and its subjective imagination as emergent properties of the mind satisfy all of the requirements of a good explanation of its event. The subjectivity of conscious experience and its imagination comprises everything in the causal history of the complete explanation of its event.

The imagination of objectivity in conscious experience generates the content of the robust patterns of the mind that are true in the natural world and in the patterns of the minds of others as generalizations. The objective imagination describes laws of nature that govern events in the universe. The imagination in the objectivity of conscious experience constitutes the understanding of a range of principles of explanation. The imagination of the objectivity of conscious experience is a mechanism for the understanding of the minds of others. The objectivity of the content and character of conscious experience comprises the generalizations and the understanding of other minds.

Simulation is the modeling of mental state understanding in the world. The simulation model of mental state understanding consists of the scalar generation of the experience of what it is like for the organism. The modeling of understanding mental states refers to the real-world properties of experience and what it is like for the organism. The organism as a biological machine generates experience in real time that is advantageous in the real world. The generation of experience as a biological machine in the real world generates patterns of mechanism and the interaction of their causal effects.

The simulation model from a biological machine contributes to the construction of a computer model of other minds and the world. The simulation model of the computer and its synthetic components comprises a reproduction of a large-scale simulation of the world (Churchland & Sejnowski, 1992). The understanding of the experience of biological organisms contributes to the construction of the computer. The construction of computer models builds the capability for real-world performance in real time that is a reproduction of experience.

The simulation model of understanding mental states in the world through a computer model may consist of a model net to receive social input and produce social output in the world through model components. The simulation of mental state understanding from a computer model refers to strings of numbers as the social input and social output of the model net. The computer model constructs in real time the content of mental states to perform a real-world output of mental state understanding. The large-scale simulation of mental state understanding in the world builds from the social model net and its construction from the computer as its synthetic components.

The simulation model of mental state understanding consists of the generation of the understanding of other minds and the world. The generation of experience of the organism reflects the subjectivity of experience in the social world. The performance of real-world output from a computer model simulation of mental state understanding is an exploration into the mechanisms of organized systems, their parts, and causal effects. The exploration in real-world output from a computer model performs a reproduction of the large-scale simulation of social minds in the world.

Simulation and Computation

The computational principles of simulation include the structural and dynamic aspects of social information-processing mechanisms. Social information-processing mechanisms demonstrate principles of computation including multiple constraint satisfaction and bidirectional network connectivity. Computational simulation is a byproduct of the patterns of activation from bidirectional social networks and their connectivity. The activation patterns of social networks demonstrate the states of functioning within bidirectional neural networks that show multiple constraint satisfaction.

The computational model of simulation consists of the structural and dynamic aspects of social networks for mental state understanding. The structural aspects of social networks consist of the specialized nets for the functional performance of the social patterns of other minds. Social computation consists of the performance of specialized social nets for the construction of output that is accurate in the social environment.

The dynamic aspects of social networks consist of the social patterns of activation that show multiple constraint satisfaction. Brain dynamics of social neural networks consist of patterns of functional activation and their connectivity. The functional activation patterns of bidirectional neural networks demonstrate a state of activation that satisfies that constraint of the environment across multiple inputs. The state of activation that satisfies multiple constraints illustrates a state of harmony in the social network. The local attractor dynamics will converge activation patterns towards an optimal state of activation for the social network.

For biological machines, social neurocomputation is a demonstration of the patterns of activation of social neural networks that generate social knowledge in the natural world. The patterns of activation of social neural networks generate a state of activation that is a response or interpretation to environmental input. Patterns of social neural network activation from social neurocomputation show the biological plausibility of computational models of social networks.

From the biological organism to the cultural computer, the understanding of mental states consists of computational components for the simulation of other minds in the social world (**Table 8.1**). At the computational level, simulation consists of a set of functional task components for the sharing and understanding of mental states. The functional specialization of social processes builds from sets of social tasks and its components at the level of computation. Social tasks perform specific functions for simulation. Computational modeling of simulation consists of functional tasks and their task components for the sharing and understanding of mental states.

Simulation consists of cognitive and affective subcomponents for the understanding of mental states across individuals and groups. Mental state understanding among individuals includes the sharing of social and emotional states across individuals for the understanding of other minds. Across individuals,

TABLE 8.1 Simulation of the mind and machine

Mind	Machine
Simulation	Synthetic
Generation	Construction
Scalar	Component
Biological	Computer
Experience	Model net
World	Output

the sharing of social and emotional states for the simulation of another mind consists of specific social and affective subcomponent tasks. The simulation of emotional events consists of a sequence of functional social and affective subcomponent tasks to produce the antecedents and consequences of social and emotional experience.

The simulation of emotional events includes the synchrony of the mental and physical states of experience and its subcomponents across individuals and groups along the spatiotemporal dimensions of the social environment. The generation of the experience of emotion arises from the perception and encoding of emotional information from social sense data. The physiological arousal and valence of social sense data as the representational content of emotion generate a network of experiential feeling states across distinct spatiotemporal scales.

The simulation of emotional events consists of the generation of experiential feeling states that may occur simultaneously or in a sequential fashion. Cultural and social patterns of feeling states that are shared across individuals and groups contribute to the individual and collective expression of emotion. Simultaneous sharing of feeling states across individuals contributes to the relational properties of social emotion. Cultural patterns of feeling states shared through simultaneous generation are reflected in the accuracy of emotional expression and its regulation. The properties of emotional expression and its regulation contribute to the generation and construction of cultural patterns in the real world and its simulation.

At the level of algorithm, simulation consists of a formal procedure for mental state understanding that produces accurate output given specific input. Simulation consists of algorithmic procedures for the production of understanding of mental states. Specific input into formal procedures of simulation consists of a set of social input. The social input of simulation algorithms includes the set of social concepts that contain the representational content of social patterns about other minds. Given particular sets of social input, simulation algorithms output social mental states such as character or situational trait attributes that describe generalizations of other minds in social patterns. The input of social concepts into simulation algorithms produce social attributions that are true in the natural world.

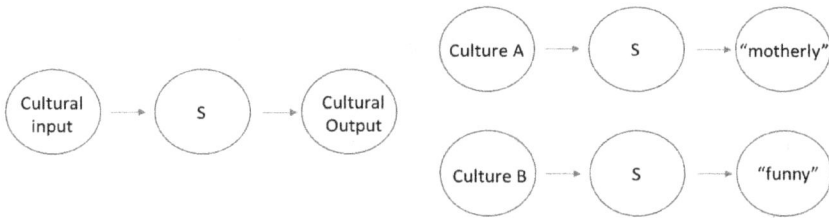

FIGURE 8.1 Culture-specific algorithm of simulation.

Culture-specific algorithms of simulation illustrate the production of trait attri-bution for distinct social patterns (**Figure 8.1**). Simulation algorithms are culture-specific such that the input and output of simulation procedures are generated for cultural systems. Social attribution as a cultural simulation algorithm depicts mental state understanding that varies across cultural contexts. The cultural algo-rithm of dispositional trait attribution refers to simulation in social patterns of the self system. The social input of a particular person or social character in a cultural simulation algorithm is necessary and sufficient to produce trait attributions that are generalizable across interactions in the social environment. The cultural algo-rithm of situational trait attribution describes simulation in social patterns from social concepts that are relational or role-based. The social input of a particular role or social relation in a cultural simulation algorithm is necessary and sufficient to produce trait attributions that are accurate and satisfactory to produce harmony across interactions in the social environment.

Computational models of simulation include a model of the physical implemen-tation of mental state understanding. Social evaluation in the self system is related to neural activation patterns in the medial prefrontal cortex and interconnected subregions of cortical midline structures. Simulation and its physical instantiation in neural networks are a form of social neurocomputation. The social information-processing mechanisms in patterns of network activation encode the representa-tional content of social sense data and its transformation into social behavior.

The cultural patterns in social neural network activation demonstrate the phys-ical instantiation of cultural processes. The physical implementation of simulation in the cultural brain consists of cultural patterns of activation in social neural networks. Cultural patterns in social neural networks consist of distinct levels of functional activation in subregions of social neural networks during social evaluation. Levels of functional activation in social neural networks depict the dynamics of brain response to social sense data in the environment. The cultural patterns in social neural networks contain the activation states that produce the optimal state of the network.

From small- to large-scale simulation, the cultural patterns of social emotion depict the advantages of real-world interaction and the simulation of the world from biological machine to cultural computer. Cultural patterns of social emotion comprise a slow simulation of the world. The cultural patterns of social emotion consist of social experiential states as a postulation of the real-world advantages from interactions in real time. For the generation of cultural patterns, the relational properties of social emotion undergo rates of change across spatiotemporal scales in the real world. The cultural properties of social emotion from biological machine are simulation through qualia and introspection. For the construction of cultural patterns, the properties of social emotion are simulated from computer through a cultural model net. The simulation of the cultural properties of social emotion from computer consists of the model net input and output.

Simulation of emotional experience as a sequence may also generate feeling states of complementarity. Complementarily of feeling states is characteristic in rates of change and in the properties of feeling states across individuals and groups. The complementarity of feeling states comprises social and cultural patterns of emotion. Cultural patterns of feeling states shared through sequential generation demonstrate the variation of emotional expression and its regulation. The relational properties of social emotion that undergo sequential rates of change generate social experiential states of distinct spatiotemporal scales. Cultural patterns of social emotion in sequential generation illustrate the subtlety of emotional accuracy and its regulation.

Culture and Simulation

Simulation contributes to the production of culture and the environment through the sharing of the representational content of other minds. The content and character of other minds consist of the representational content of the social environment in the natural world. The generation and sharing of mental states across minds comprise the social processes for the production of cultural variation in the social environment.

Cultural variation in the social environment is produced from the generation and construction of cultural patterns of other minds in interaction with the natural world. Cultural patterns refer to the generalizations of the natural world that are true and reflect the understanding of other minds. Cultural patterns consist of the social processes that contribute to the production of cultural values, practices, and beliefs. The simulation of cultural patterns through social learning is a means for the acquisition of mental state understanding.

Cultural values, practices, and beliefs consist of the cultural patterns that comprise the societal structure of social experience. Cultural patterns as valuation states are defined through the experience of social content and social character. Cultural patterns as behavioral states show the real-world properties of experience through the practice of social content and social character in interaction with the social environment and the natural world. Cultural patterns as belief states are comprised of regularities of conscious experience in patterns of nature that hold.

Cultural dimensions consist of the set of social processes that consist of mental and physical property of other minds. The mental and physical property defined by cultural dimensions is generated from the knowledge of self and others. Cultural dimensions guide the representational content of knowledge from self and others in the social environment. Cultural dimensions assume that the generation of social knowledge arises from the sharing of mental states across minds. The generation of social knowledge about the self arises from the sharing of mental states of the self with others, while the sharing of mental states of others with the self constitutes the acquisition of the social knowledge of others.

The mental and computational property of mental state understanding is necessary for the generation of the causal attribution of social phenomena in the natural world. The interaction of social phenomena and their causal effects constitutes a componential feature of the social environment. The attribution of mental state in social phenomena comprises a mechanism of causal explanation in the social environment. Other minds as social phenomena consist of the generation and construction of mental states through interaction with the world. The trait attribution of mental states in another mind assumes representational content of mental states that is consistent with characteristics. The situation-based attribution of mental states in another mind assumes representational content of mental states that is defined in the context of the social environment.

The simulation of understanding other minds encompasses experiences of understanding social relations and their causal effects in the social environment and the natural world. The experience of social trait attribution is feature of the social environment with causal effect. Social trait attribution as a feature of the social environment consists of the character trait as a social concept. Social trait attribution is a causal projection of the mental states of other minds as a generalization. The simulation of social trait attribution is a construction of states of social knowledge of other minds independent of the social environment. The attribution of social traits as a generalization is a prediction of social patterns and their causal effects.

The experience of situation-based attribution is to recognize a social mechanism defined as a configuration of parts with causal effect. The configuration of parts as a situation-based attribution is the arrangement in the environment of the scene and its objects as the parts. Situation-based attribution is a causal projection of the mental states of other minds within a future context. The attribution of social traits based on a situational context is a prediction of social patterns and their causal effects within a historical context. Higher-level features that arise from a configuration of parts with causal effects may relate to the causal effect of situational trait attribution. The simulation of situation-based attribution is a generation of states of social knowledge of other minds in interaction with the social environment. Thus, the simulation of other minds in a social pattern consists of a complex set of causal effects within a social system.

The mental simulation of social attribution in biological organisms reflects an emergent property based on experience. The experience of social attribution is a social pattern for the recognition of causal effects within a social system. In the social environment, social attribution generates the social experience of causation from other minds. In the natural world, social attribution generates the social experience of causation from the understanding of the mind of others.

The organism as a biological machine generates the experience of social attribution in the social environment and the natural world. The generation of the experience of social attribution to other minds in the biological machine is a causal effect without explanation at a lower level of analysis. In the social environment, the conscious experience of social attribution as a social pattern is a recognition of causal effect within the self system. In the natural world, the conscious experience of social attribution as a social pattern is a recognition of causal effect within the social system.

The experience of the organism as a biological machine generates social patterns and their causal effects. For the biological machine, social patterns and their causal effects demonstrate real-world advantages. Understanding other minds generates real-world properties of experience that confer advantages. The understanding of other minds builds social knowledge of other minds in interaction with the social environment. Social patterns in the real world are generated from the self system in the social environment. Social knowledge of self and others contributes to the social patterns that demonstrate causal effect in the real world.

Mental state understanding in the real world is advantageous for biological and cultural adaptation in real time. The simulation of other minds in the real world demonstrates social patterns that unfold in real time. Mental simulation in real time is a construction of social patterns through the generation of social knowledge in the real world. For the biological machine, mental simulation in real time is an expression of the advantages of the real-world properties of experience.

The simulation of other minds contributes to the social processes of cultural patterns. Understanding mental states generates the experience of ingroup social recognition that is consistent with patterns of cultural selection. The accurate recognition of the mental states of ingroup members is an advantage of cultural group selection. Understanding mental states is a mechanism for the content biases in cultural patterns. The preferential encoding of social information because of its representational content in the mental states of others reflects the content biases in cultural patterns. The act of simulation is a causal mechanism for biasing the content of social information into cultural patterns. Mental state understanding in intergroup contexts is a causal mechanism for the generation of social information into patterns of cultural variation.

Simulation in Culture

Culture and the environment contribute to the generation and construction of the societal structure and the social patterns of other minds. In response to environmental demands, cultural processes consist of the societal structure and social patterns that constitute the content of other minds across spatiotemporal dimensions. Cultural processes define the societal structure and mental content of social relations that constitute the social environment. The simulation of the mental content of other minds contributes to the cultural process.

Cultural evolutionary processes refer to cultural biases that demonstrate an advantage for the cultural group in the real world. Cultural biases reflect the generation and transmission of cultural content through specific social mechanisms (Mesoudi, 2009). Cultural content biases refer to the spread of social thought through specific cultural content. Cultural model-based biases reflect the transmission of cultural content through social learning from a cultural model. Cultural frequency biases include the cultural transmission of social thought because of the popularity of its content in the population.

Simulation as the understanding of other minds is a mechanism of the social environment that contributes to the cultural evolutionary process. Through simulation, the mental content of other minds generates social knowledge that is advantageous for the cultural group and responsive to environmental demands. The simulation of the mental content of cultural processes in other minds demonstrates the generation of social knowledge for cultural transmission. The simulation of mental content from a cultural model to learner is the generation of social knowledge for cultural transmission through social learning. The simulation of the mental content in the population reflects the cultural transmission of social knowledge through popularity.

Cultural niche construction reflects the coevolutionary processes of dual inheritance theory. The cultural niche construction of the environment includes cultural changes in the ecological niche and its inheritance that are adaptive for the organism. Cultural changes in the ecological niche through cultural processes are a mechanism of cultural and biological adaptation. The generation of cultural processes from the organism constitutes the emergent properties of cultural and social patterns that arise in interaction with the social environment and the natural world. Through cultural changes, the ecological niche and its inheritance demonstrate the patterns of cumulative cultural evolution across generations of social learning.

Environment conditions may lead to the generation and construction of patterns of cultural and social thought as a causal mechanism. The societal perception of threat in the environment is a causal mechanism for protection. Cultural and social patterns act as a protection to environmental threat through its prevention and intervention. The simulation of cultural and social patterns of thought is a source of prevention and intervention to environmental threat.

The generation of societal perception of safety and protection is important for the prevention to environmental threat. Cultural patterns serve as a social resource of prevention through the generation of the societal perception of protection. Cultural patterns of social thought consist of content biases in social information that are advantageous and protective. Cultural patterns of social thought generate mental content that bolsters societal perceptions. The simulation of cultural patterns of social thought is a strategy of prevention to specific environmental threats.

Cultural patterns as a social buffer act as an intervention through the construction of societal conditions that provide the necessary and fundamental resources for protection. Cultural patterns of social thought promote the societal conditions important for social development and empowerment. The construction of societal conditions for cultural and social patterns of equality is fundamental to the strategies of social development and empowerment.

Mental Simulation

The mental content of simulation arises from the social constructs of multilevel processes. Understanding other minds generates social knowledge at the level of the individual and is a component of relational social knowledge at the interpersonal level. Simulation as a social mechanism is foundational to social cohesion and social cooperation at the intergroup level. The mental content of simulation is deterministic through interaction with the social environment.

Social knowledge and its generation consist of the knowledge of self and others. The social knowledge of self and others is stored as a hierarchical structure with interconnected branches to social mental content. Mental content of the self and others is comprised of social concepts including trait attributes that describe social and emotional states of the mind. At the individual level, the social and emotional states of the mind are thought to consist of generalizations that are consistent across spatiotemporal conditions.

Relational social knowledge is comprised of the concepts of self and others that are defined through social roles and relations. Fundamental social relations depict distinct relations of the self and others defined along a continuum of equality. Relational social knowledge consists of the knowledge of self and others that is defined in relation to others through social roles. Social roles consist of the mental content of self and others for the performance of shared goals. Social relations connote the mental content of self and others that consists of situational attributes or social and emotional states of the relational mind.

The relational mind is focused on the generation and sharing of mental states that orient the mind towards salient goals. Relational mental states consist of social emotions characterized along the dimensions of valence and arousal and their antecedent in the self and the social environment. The relational mind produces social and emotional states consistently towards the attainment of the relational or the shared goal. Relational mental states consist of social patterns of mental thought towards a state of harmony.

Intergroup social knowledge refers to the conceptualizations of self and others that are defined through the social group. The identification of the self with the social group is foundational to intergroup social and emotional states of belonging and commitment. The mental content of social identity is defined as the knowledge of self and others as defined by the social group and is important to the conceptual formation of self identity and social belonging. Social group membership consists of the mental content of ingroup recognition that is advantageous to group members.

The intergroup mind produces social and emotional states that promote social cohesion and social cooperation. The mental content of the intergroup mind consists of social information that is advantageous to the social group and promotes intergroup relations. The mental content of the intergroup mind promotes relational social knowledge that consists of social patterns that are harmonious across groups. The intergroup mind generates and shares social patterns of thought that guide the activity of groups towards a state of satisfaction or harmony.

Culture and Mental Simulation

Culture promotes the understanding of other minds through the generation and construction of cultural patterns of social thought. Cultural variation in the social mind constitutes a source of variation in the natural world. Cultural processes contribute to societal structure and social relations through the satisfaction generated from the understanding of others. Cultural patterns of social thought consist of the mental and physical property that promotes states of harmony of the individual and group.

Simulation guides social nets and its activation patterns towards the optimal states for production of cultural processes. Computational principles of simulation depict the structural and dynamical aspects of the social nets and their activation patterns. Social computation in patterns of network activation encodes the representational content from sense data into hidden layers of interpretation to produce output from the social nets. Social neurocomputation consists of the activation patterns in social neural networks that generate states of social experience of the organism through its interaction with the natural world.

For biological machines, mental simulation is a form of social neurocomputation. The mental content of simulation as the understanding of mental states arises from the generation of the states of social experience from other minds. Simulation consists of the understanding from the mental and physical property of the individual and the group. The representational content of the individual and group in the social environment, its mental and physical property, comprises the social sense data of the social net.

The mental simulation of biological organisms is an emergent property based on experience in the real world. Biological organisms are the generator of emergent cultural phenomena in the natural world. The mental content of its simulation depicts the cultural phenomena in the natural world that is emergent from the mind. The mental content of emergent cultural phenomena arises from the patterns of activation of cultural neural networks. Simulation of the mental content of other minds is a mechanism for understanding the emergent cultural phenomena of biological organisms.

For cultural computers, simulation is a form of social computation. Computational modeling of simulation is a construction of the world from a model net. The model net and its component parts construct the mental and computational property in the real world for simulation. The rates of change of mental and computational properties from the model net and its component parts reflect production of social performance for real-world advantages in real time.

The simulation of cultural computers is a synthetic production of mental and physical properties for advantages in the real world. Cultural computers are the constructor of cultural property in the environment. The simulation of cultural model nets and their component parts illustrate the construction of real-world properties within spatial and temporal dimensions. Cultural model nets depict the social performance of real-world properties and their rate of change for advantages in real time that are optimal for the cultural computer.

Mental state understanding consists of the mental and physical properties that define distinct cultures. Cultural variation in the natural world arises from the mental and physical property. Cultural patterns of social thought in the natural world describe the mental and physical property of the biological organism that are generated to strengthen the optimal state of cultural processes. The conscious experience of what it is like is an emergent property of the biological organism. The generation of cultural patterns in the biological organism reflects the mental and physical properties that define optimal states of cultural processes.

Understanding the mental states of other minds is a real-world advantage of cultural processes. Simulation is a construction of the social performances in the real world that demonstrates advantages in real time. Mental state understanding from the cultural model net produces rates of change in properties that are consistent with the optimal states of the model net. The cultural model net consists of the mental and physical property of simulation and produces interpretation of such input for real-world advantages.

Distinct cultural processes consist of social knowledge that leads to cultural and psychological adaptation. The understanding of mental states across distinct cultures is fundamental to cultural and psychological adaptation. Culture-specific algorithms produce social knowledge that is deterministic and accurate within a given culture. Across cultures, mental state understanding requires the acquisition of social knowledge that is beneficial for distinct cultures and environments.

Across environments, simulation in cultural model nets facilitates the construction of cultural products for real-world advantages. In digital environments, cultural model nets simulate mental content for the learning algorithms of biological organisms. In physical environments, cultural model nets construct mental content from cultural model nets that produce an optimal state of satisfaction for cultural processes. From digital to physical environments, cultural processes guide the construction of mental content for the understanding of other minds.

Conclusion

Understanding other minds is an essential mechanism of the cultural mind. The cultural mind is understood through the simulation of its mental content. The cultural mind is comprised of the mental content that represents cultural phenomena in the natural world. The mental content of the cultural mind consists of mental and physical property of cultural phenomena that exists in biological organisms and their interaction with the natural world. The mental content of the cultural mind is comprised of the mental and physical property that is constructed through computational models of cultural phenomena and their underlying processes.

The notion of the cultural mind as a cultural brain is a generalization of the emergent cultural phenomena in the biological organism. The mental content of the cultural mind corresponds to the physical instantiation of cultural phenomena in the cultural brain. The mental events of the cultural mind are related to physical events of the cultural brain. The simulation of social thought is a mental event of the cultural mind that corresponds to a physical event in the patterns of activation in social neural networks of the cultural brain.

The generalization of emergent cultural phenomena demonstrates the lawlike regularity of biological organisms. The physical property of the cultural brain of biological organisms is necessary and sufficient for the lawlike regularity of emergent cultural phenomena from the cultural mind. The physical events of the cultural brain as the patterns of activation from social neural networks are a mechanism for the experience of mental content that is cultural. The cultural mind as a physical instantiation of patterns of neural activation demonstrates its biological plausibility.

The notion of the cultural mind as a cultural computer is a computational model of the cultural mind. The cultural mind as a computational model consists of the mental content of the cultural mind for the construction of cultural phenomena in the environment from synthetic parts and devices. Simulation as a mental capacity of the cultural mind is a function that can be modeled to perform through a synthetic part or device. Simulation as a social performance consists of the construction of mental events through the output of computer devices. The cultural mind as a computational construction demonstrates the capabilities of social performance that are optimal for the cultural model net.

The cultural mind as a cultural computer is a form of niche construction. The performance capability of cultural computers to understand other minds is a demonstration of real-world advantages in real time. The simulation of cultural computers is based on the mental content of the cultural mind in the natural world. The simulation of a cultural computer refers to the understanding of other minds from the cultural construction of mental and physical property across environments. The construction of mental and physical property across environments is a demonstration of the mental content of the cultural mind and a physical instantiation of cultural phenomena in the world.

References

Churchland, P.S. & Sejnowski, T.J. (1992). *The computational brain.* Cambridge, MA: MIT Press.
Mesoudi, A. (2009). How cultural evolutionary theory can inform social psychology, and vice versa. *Psychological Review, 116(4),* 929–952.

PART III

9
ARTIFICIALISM

Introduction

Artificialism is a philosophical inquiry into the relation of computation to the mind and the role of scientific structure in computation. As a philosophical approach, artificialism explores the nature of mind and its role in possible worlds. Artificialism implies a role for thought as the mental content that comprises the continuity across possible worlds. The computing mind produces mental content as the part and particular of mental life from physical to virtual worlds.

The computation of minds and machines introduces a range of philosophical questions about the nature of mental content and its role in the cultural sphere. The establishment of a standard in performance and production of informational content enriches mental life. Cultural life is broadened through the expansion of thought and reason as a reflection of truth in the cultural sphere. The mental content of cultural life is a source of knowledge of discovery. Artificialism entails the intentionality of intelligent design in mental life. Artificialism explores the role of computation in philosophy of mind and philosophy of science.

Artificialism

Artificialism as a philosophical approach refers to the relation of computation in computer science to philosophy of science. One role of computer science in philosophy of science is for the purpose of simulation and construction of parts and particulars of living systems through computation. Another role of computer science in philosophy of science is for the simulation and construction of parts and particulars of living systems in technology such as devices and computers and their related programs.

The science of computation contributes to the understanding of the mind as mental computation and information production as machine computation. Understanding the mind as mental computation in possible worlds informs the epistemological foundations of the nature of thought in philosophy of mind and philosophy of science. The information production from machine computation establishes the capability for the reconstruction of parts and particulars of living systems in possible worlds as a standard in philosophy of mind and philosophy of science.

Artificialism entails the relation of theories in computer science and theories in philosophy of science. Artificialism articulates the implications of the principles of computation for philosophy of mind and philosophy of science. Theories in computer science describe the principles of computation that guide the design and construction of programs and computers of possible worlds. Computational principles describe the design and construction of programs and computers in the living systems of possible worlds. Computational discoveries inform core considerations of the importance of unbiased information production in computational discovery, the determination of valuation in computational performance, and the pertinence of the computing mind in philosophy of mind.

Varieties of artificialism range from questions of computation as a science to philosophical inquiry as computational scientific discovery. Artificialists consider the standards of accuracy, reliability, and efficiency of information production in computational theory and design as foundational to theory building in philosophy of science. The information production from mental and machine computation as computational discovery contributes to the neutrality of computational theory.

Computer science as a science is a resource for information production and computational discovery in philosophical inquiry, rather than a replacement of philosophy. Information production as computational discovery is the production of data for the design of policy. Computational scientific discovery informs the production of policy design. Artificialism supports discovery science and its contribution to philosophy of science through computational discovery processes.

The computing mind consists of parts and particulars that perform as machine property in states of machine computation. Machine property is the property based on machine parts and its configuration for the performance of tasks. Machine property is the property of machine parts and its configuration for computation and information production. Machine property consists of the machine states that perform information production based on programmable rules or instructions. Machine property comprises the machine parts and particulars that produce machine output based on a programmable machine table of preset instructions. Machine computation produces deterministic output based on a machine table with a given particular input.

Machine capital is the valuation and worth of machine property. Machine capital entails the valuation and worth of machine property based on task performance. Machine capital is the valuation and worth of machine property based on computational performance and information production. The automation and control of information production as machine property entail the performance of mental and machine computation. The determination of valuation in computational performance is based on the protection of machine capital. Protection of machine capital is defined as the standard of accuracy, reliability, and efficiency of function and performance in information production and its automation and control.

The machine equivalence of information production based on different rules implies distinct machine identity. Machine identity as a scientific and technological concept connotes the machine state or machine entity and its machine property. The construction of the machine state entails the protection of its machine property and machine capital through its reconstruction and replication. The machine state consists of the parts for performance equivalence and information production that are autonomous and self-organizing across levels of a living system. The machine equivalence in performance is not necessarily the same machine computation. The machine equivalence in performance of functional roles demonstrates the importance of machine functionalism. The protection of machine capital implies a standard of machine equivalence in performance and information production.

Artificialism as a philosophical position acknowledges the requirements for theory building in computer science as a distinction from the standard normal science paradigms of empiricism. Theory building in computer science is based on the simulation and construction of computational models as programs. Computational models are formal models that provide a tool for the construction and testing of causal explanations in datasets. Building programs for computational modeling and data fitting to models allows for the testing of formal hypotheses. Computational modeling is essential to theory building in computation as it allows for the testing of formal hypotheses and the formalization of theoretical assumptions into unified frameworks.

Artificialism conceptualizes the boundaries of philosophical inquiry in computer science as theory building about the computation level of biological systems. Philosophical positions in artificialism define the criteria of equivalence in machine performance and the computational properties of organized systems that are consistent across possible worlds. Philosophical inquiry into the parts and particulars of the science of computation explores a delineation into the role of scientific and technological progression.

The expansion of scientific computation from biological into synthetic devices connotates scientific and technological progression. The advancement of computation as a science introduces distinct terminology and novel research programs that explain genuine representations in the structure of the world through the lens of scientific and technological realism. The fundamentals of computational discovery science advance novel methods of observation and approaches to theory building as a scientific tradition.

The computing mind describes a standard of mentality and intelligence in philosophy of mind. Artificial intelligence describes the performance of functional tasks from machine computation and the production of informational content from the computing machine. Artificial intelligence is the production of informational content based on the machine computation of specific functional roles. Artificial intelligence demonstrates the capability of computers for machine computation that shows a functional equivalence to mental computation. Artificial intelligence contributes to the standard of mentality of the computing mind, and the understanding of the functional role of intelligence in mental life.

Artificial neural networks describe the computational modeling of neural networks. Artificial neural networks describe the abstract formal modeling of activation states of neural networks that are simple approximations of the biological mechanism. The approximation of the activation states of neural networks in computational models describes formalizations of the computational properties of neural networks. Artificial neural networks imply a functional role of neural networks in mental and machine computation. Computational models of artificial neural networks test the biological plausibility of neural networks.

Artificial life describes the computational properties of artificial life as a living system. Artificial life as a model of the features of living systems is comprised of the computational properties that are autonomous and self-organizing (Godfrey-Smith, 2003). The artificial life system consists of cellular automatons and interacting elements into self-sustaining patterns. Artificial life depicts the design of programs in formal models as the fundamental properties and patterns of living systems. Artificial life demonstrates the capability of programs for the construction and testing of formal models on the properties and patterns of living systems.

Artificialism and Computation

The design and construction of programs and computers entail the simulation and construction of the physical world into digital and virtual worlds as possible worlds. The physical world as a possible world entails philosophical notions of physicalism from nonreductive to supervenience physicalism. The computational principles that describe the physical parts and the functional elements of the biological computing machine consist of the spatiotemporal properties of mental computation. The computational models of the biological computing machine describe computational principles that guide mental computation.

The biological computing machine that produces mental computation entails the structural and functional organization of the brain. The structural and functional organization of the nervous system describes the levels of processing and biophysical mechanisms that transform representational content into motor output. The biophysical mechanisms of the nervous system transform sense data into motor output based on computational principles across levels of functional organization. The biological computing machine defines a standard of mentality and intelligence of the computing mind.

The computational components of artificial neural networks consist of models that describe simplified approximations of the bidirectional connectivity of activation states in neural networks. The computational models of neural networks depict the activation patterns of neural networks and their causal relations. Computational models of neural networks identify the causal relations from the properties of neural networks.

Artificial neural networks consist of the simplified approximations of patterns of activations from the properties of neural networks. Artificial neural networks are comprised of the net functions that describe the sense data as weighted input of a net of activation states to produce an output. The computational components of artificial neural networks consist of net functions that can be built into machine properties, from integrated circuits and artificial neural structures to synthetic devices and technology.

Artificial intelligence as machine computation of devices and computers consists of the programs and procedures that perform and produce informational content. The computational components of early machine computation consist of the physical states and their transitions that are rule-based and their functional operations. Rule-based programming in machine computation consists of the machine table as programs that define the functional operations as procedures of performance and information production.

Artificialism is a computational approach to the formal modeling of the properties and parts of artificial living systems. Artificial living systems describe the formal modeling of features of living systems as autonomous and self-organizing computational properties. Artificial life as a self-organizing living system describes the autonomous dynamics of information flow. Artificial life is comprised of cellular automata that perform self-replicating programs as part of the living system.

Theory building in artificial life consists of the simulation and construction of models as programs. Theory building in artificial life is comprised of the construction and testing of causal explanations from models and data resources. The data extraction and replication from knowledge-based resources contribute to data fitting as testing of theoretical models.

Artificial life comprises the spatiotemporal properties of possible worlds from physical to digital world and physical to virtual world interactions. The properties of the physical world consist of multiple physical realizers such as the environments of digital and virtual worlds. The simulation and construction of real-world experience of the physical world and its interaction are built into the digital environment and immersive virtual environments. The real-world experience of digital and virtual environments comprises a reality that consists of a model of representations that is consistent across possible worlds. Digital and virtual identities and their groups protect the interactive features of digital and virtual environments through autonomous self-organization. The net-to-net interactions of digital and virtual environments perform as interfaces of real-world experience in possible worlds. The real-world experience of artificial living systems performs as a unified, biological machine.

Artificialism and Culture

Artificialism considers the impact of computation on culture. The computation of minds and machines comprises a source of knowledge generation and information production of living systems. The cultural invention of devices and computers as machine computation demonstrates the use of tools for the production of knowledge and information. The machine computation of knowledge-based systems and information production contributes to the protection of cultural capital.

The design and construction of cultural programs and cultural computers demonstrate the capabilities for cultural development and cultural advancement. The design and implementation of cultural programs consist of the dissemination of resources for the benefit of cultural development. The cultural participation in activities and the attainment of program goals further advancement in the cultural and public sphere. The development and implementation of cultural programs assist in the maintenance of cultural traditions and the equality of relations among diverse cultural groups.

Cultural programs inform the production of cultural data for the design of policy. Computational scientific discovery informs the production of policy design in the cultural and public sphere. Computational scientific discovery produces data resources for the design of evidence-based policy at the cultural level. The use of computational scientific discovery for the production of the information cycle contributes to the implementation of policy design, from problem identification and option development to decision making and evaluation of policy making and governance.

Computation shows how scientific and technological progress informs cultural development. Scientific and technological progress contributes to the design of tools for the building of initiatives and programs that disseminate information to the community. The advancement of technological approaches for community building and cultural participation is a resource for the benefit of cultural development. Scientific and technological approaches impact the design of tools and resources for economic and social empowerment. The implementation of programs for economic and social empowerment contributes to development in the cultural and public sphere.

Artificialism is an advancement of computational discovery science in the cultural and public sphere and its contribution to philosophy of science. Cultural programs contribute to the information production of computational discovery. The determination of valuation in computational performance of cultural programs is based on accuracy, reliability, and efficiency as a standard of performance. Computational discovery science sets standards of performance that demonstrate the pertinence of the cultural computing mind in philosophy of mind.

Cultural machine property consists of the states of machine computation at the cultural system level. Cultural machine property is the property based on machine parts and its configuration for performance and information production. Cultural machine property consists of the machine states that perform information production based on programmable rules.

Cultural machine capital is the valuation and worth of cultural machine property as the machine property at the cultural system level. Cultural machine capital is the valuation and worth of machine property based on computational performance and information production at the cultural system level. The automation and control of information production as machine property entail the performance of computation at the cultural system level. The protection of cultural machine capital consists of the reconstruction and replication of cultural property.

The protection of cultural machine capital as the standard in function and performance is the criterion for determination of valuation in computational performance. The protection of cultural machine capital implies a standard of performance equivalence in information production at the cultural system level. The standard of cultural machine equivalence in performance and information production implies the protection of cultural machine capital.

Theory building in computer science consists of the simulation and construction of cultural programs. The simulation and construction of cultural programs in computer science provide a tool for the formulation of theoretical assumptions into unified frameworks. The design and testing of cultural programs as computational models allow for the building of devices and computers for cultural performance.

The building of devices and computers for cultural performance consists of the design of cultural programs. Cultural programs seek to simulate the performance of cultural tasks that are consistent with those of natural phenomena. Cultural programs that perform cultural computation include programs designed for the production of artificial language. Computer programs that produce artificial language demonstrate the features of artificial language production from user-generated scripts. Artificial language production follows the rule-based structures of natural languages.

Artificial language production is an example of the application of artificial intelligence for purposes of social communication in living systems at the cultural system level. Artificial language production demonstrates the biological plausibility of neurolinguistic mechanisms. The use of artificial language production in social communication is the production of machine computation for mental computation. Artificial language is a tool that assists in the production of social communication.

Computational approaches to cultural neuroscience contribute to the design of computational models that discover the patterns and mechanisms of cultural neural networks. Computational models of cultural neural networks contribute to the knowledge generation of cultural processes in the social and physical world. Computational models simulate and explore the cultural mind. Understanding the computational level of the cultural mind contributes to the representation of the cultural computing mind and the mechanisms of cultural computation.

Artificial life further demonstrates the role of artificial intelligence in the functionality of the cultural system level. Artificial life shows how the performance of machine computation as autonomous and self-organized advances the simulation and construction of artificial living systems at the level of the cultural system. The part and particulars of the machine property of artificial life contribute to cultural property. The machine capital of artificial living systems is part of the organized system of cultural capital. The conceptualization of artificial living systems represents a demarcation of artificialism from naturalism. Artificial living systems are a source of knowledge generation that underscores the significance of reconstruction and replication as complementary processes to reductionism in cultural systems.

Culture in Artificialism

Artificialism supports a culture of computational discovery science and its processes. The culture of artificialism consists of the design and application of technology for the use of knowledge-based resources. The culture of technology educates the scientific mind in computational science and discovery science. Formal science education trains the mind for observation of the physical world and the design of technological tools for theory building.

Cultural advancement as scientific and technological progress is the expansion of understanding of the cultural sphere through the design of tools for the simulation and construction of cultural life and cultural traditions. Technological tools designed to enhance the understanding of the importance of cultural diversity, inclusion, and equity in the public sphere contribute to cultural advancement. Technology builds awareness of the societal benefits of cultural diversity, inclusion, and equity of ethnic and cultural groups.

Technological progress provides resources for the discovery of the cultural sphere as a way of life. The cultural sphere consists of the multilevel functions for the automation and control of cultural production. The cultural development of technological tools consists of the reconstruction of cultural models as social representations. Technological tools allow for the growth of knowledge of social representation in mental life and in possible worlds. The use of technological tools for learning and education provides cultural resources for knowledge acquisition.

The use of technological tools for cultural advancement includes the use of knowledge-based resources as a scientific and technological resource. Cultural programs that perform cultural computation are designed to assist the cultural production in the artificial living system. The design of cultural machines for the performance of cultural computation constitutes functional equivalence in the production of mental computation to machine computation at the cultural system level. The use of cultural computers to assist the automation and control of cultural production is the performance equivalence of machine computation to machine computation. The automatous cultural dynamics of the artificial living system are a resource for the performance of machine computation for mental computation.

Cultural development as scientific and technological progression is the building of knowledge-based resources for the automation and control of cultural production. Programs in cultural development support activities and participation in cultural traditions. Cultural development programs contribute to the societal perception of equality of ethnic and cultural groups and the societal benefits of inclusion and cultural diversity. Programs in cultural development support social and economic development in the cultural and public sphere. Scientific and technological progression contributes to the promotion of the automation and control of cultural programs and cultural production.

Culture as scientific and technological realism is the continuity of thought and reason from the cultural to the public sphere. The cultural sphere consists of the accurate representations of the world in science and technology. Participation in cultural life is broadened with scientific and technological progression. Cultural advancement as the attainment of societal goals is accomplished through the achievements of science and technology as a truthful correspondence of the structure of the world.

Artificialism and Cultural Computation

Cultural advancement contributes to the societal perception of equality among ethnic and cultural groups. The maintenance of cultural tradition and the promotion of social relations of ethnic and cultural groups present a multitude of societal benefits. Cultural advancement builds from an understanding and appreciation of multicultural ideology and the efficacy of acculturation strategies across individuals and groups. Cultural advancement contributes to the design of national policy and the attainment of societal goals across nations.

Cultural development as the automation and control of cultural production facilitates the societal benefit of scientific and technological tools for individuals and groups. Cultural development supports the design and implementation of initiatives and programs that set the standards of automation and control in cultural production. Cultural programs and initiatives that support the development of scientific and technological tools broaden the participation of individuals and groups in the cultural and public sphere. Cultural programs and initiatives that support scientific and technological innovation build the resources of individuals and groups for its automation and control of cultural production.

Cultural programs facilitate scientific and technological progress through the access to resources for individuals and groups. The access to scientific and technological resources promotes cultural knowledge of the structure of the world. Cultural programs promote standards that support an appreciation for societal pluralism and the importance of cultural diversity, inclusion, and equity of groups. Scientific and technological tools contribute to the development of appreciation for societal pluralism and the importance of cultural diversity, inclusion, and equity. Cultural programs provide access to scientific and technological resources and tools that promote cultural advancement. The cultural development and implementation of initiatives and programs support scientific and technological innovation that meets fundamental standards of cultural advancement.

Culture consists of the fundamental elements for the automation and control of cultural production. From symbols to practices, the automated production of fundamental elements of culture serves as a model of social representation. Cultural dimensions define a standard in the automation and control of information production at the level of cultural systems. Cultural dimensions of social knowledge provide a source of sense data and mental content for cultural production and output. Cultural dimensions define the spatiotemporal properties of cultural content as mental content. Computational modeling of culture builds knowledge of the regularities and patterns in cultural nets and cultural neural networks. Bidirectional cultural neural nets describe computational components of the information-processing mechanisms of culture.

Culture as a state of consciousness contributes to the automation and control of thought and reason. The mental property of consciousness as a part of mental computation supervenes on physical property. The machine property as part of artificial intelligence and machine computation is nonreductive. The supervenience of consciousness on physical property implies that states of consciousness have computational components that are a natural source of cultural production. Cultural property as states of consciousness constitutes the influence of culture on societal perception and the cultural patterns of thought.

Conclusion

The continuity of thought from naturalism to artificialism represents a growth of knowledge from scientific and technological progression. As a scientific tradition, computational science advances novel approaches to knowledge generation and scientific discovery. The science of computation contributes to the standards of scientific social structure. Artificialism supports discovery science and its contribution to philosophy of mind and philosophy of science.

Reference

Godfrey-Smith, P. (2003). *Theory and reality: An introduction to the philosophy of science.* London: University of Chicago Press.

10
MACHINE LEARNING

Introduction

Machine learning in philosophy of mind explores the role of change in mentality and thought. Machine learning implies that the computation of biological and synthetic machines undergoes physical state transitions that perform a functional role. The change in mentality and thought is a cooccurrence with the change in physical property. Computational principles explain regularities in patterns of thought in mental content and in the structure of the world. Changes in the directionality of patterns of thought and mental content are understood through computational principles of interactivity.

Machine learning as a causal power of the mind describes the causal role of knowledge generation on patterns of thought and their physical realizers. Machine learning of minds plays a causal role in the generation of novel thought patterns and their physical implementation. Machine learning of computers demonstrates the causal role of knowledge extraction on patterns and regularities in the structure of the world and its interactivity. Learning algorithms that perform data fitting to a model reinforce patterns and regularities from the input.

Machine learning as first-person knowledge consists of the knowledge generation from subjective experience. Real-world experience as a continuous stream of sensation comprises the input of mental computation. The spatiotemporal properties of the mental states of inference from real-world experience constitute the mental and physical property of knowledge generation. Machine learning as third-person knowledge is the knowledge generation from objective experience. The knowledge generation from objective experience comprises the mental performance from learning rules.

Machine learning as a mental and physical property describes the change of spatiotemporal properties from the computation of minds and machines. Machine learning entails the change in the spatiotemporal properties of physical states of mental and machine computation. Machine tables as learning rules describe the transformation of machine input to machine output through the transitions of physical states. The physical states and their transitions of learning in the biological computing machine are depicted as biophysical mechanisms in the structural and functional organization of the nervous system. Computational models of neural networks allow for the extraction of patterns of information within hidden layers.

In philosophy of science, machine learning refers to the functional role of inference in theory confirmation and evidence. The functional role of inference from minds and machines is explanatory inference as inference to the complete explanation. Knowledge generation from mental computation aims to perform theory confirmation based on reasoning. Knowledge extraction from machine computation attempts to align observation with reliable patterns and regularities in the structure of the world. Learning algorithms detect patterns in the structure of the world that inform observation and true belief. The alignment of observation with reliable patterns and regularities is the mental and physical property from interaction of the mind in the world.

Machine Learning

Machine learning refers to the notion of the brain as an inference machine. Machine learning consists of the conceptual and algorithmic tools for the computational modeling and prediction of mental inferences. Machine learning approaches to brain dynamics consist of the computational models and algorithms for the prediction of inferences of the brain (**Table 10.1**). As computational approaches to data analysis, machine learning algorithms describe the formal procedure for the prediction of inferences based on sense data at the computational level. Machine learning algorithms that predict inferences of the brain describe the computational modeling of the brain dynamics of neural networks.

Learning as a computational principle describes a change in neural plasticity that accompanies a change of growth in knowledge generation. The Hebbian

TABLE 10.1 Machine learning as computation

	Mind	*Machine*
Computational	Inference	Program
Algorithm	Hebbian learning	Deep learning
Physical implementation	Biological	Computer

learning principle describes the change in synaptic strength of neurons based on growth (Hebb, 1949). Learning is a mental construct that is physically instantiated as neural mechanisms of memory and principles of neural networks. Learning consists of biophysical mechanisms of plasticity across the levels of processing in the structural and functional organization of the nervous system. From cortical systems to molecular mechanisms, the biophysical mechanisms of plasticity shape the functional architecture of the nervous system.

Learning is comprised of the biophysical mechanisms of plasticity across levels of analysis. The computational level of learning is considered a part of the classification of learning and memory (Churchland & Sejnowski, 1992). The algorithmic level of learning consists of a multitude of learning algorithms from classical conditioning to developmental processes of neuronal maturation. The physical implementation of learning is comprised of a range of information-processing mechanisms from molecular and cellular mechanisms to functional neural networks within the neuroanatomical structures of the medial temporal lobe and interconnected brain regions.

In a broader sense, machine learning refers to knowledge extraction from large datasets. Machine learning algorithms are programs that perform data identification, such as data extraction of learning rules from patterns in datasets. Machine learning algorithms perform classification and pattern recognition based on the input of datasets. The Bayesian brain describes a computational approach to the prediction of inference from encoding in patterns of neural activation or the patterns of structural or functional connectivity across structures of the brain.

The physical states of machine learning in the computer are defined as programmable functions. Machine learning describes the learning algorithms designed for knowledge extraction from large datasets for the purpose of prediction. Computer programs that perform machine learning with learning algorithms extract patterns of information from large datasets. The functional role of machine learning in the computer demonstrates the biological plausibility of neural plasticity in the biological organism.

The interaction of biological to biological computation implies the knowledge generation of informational content for theory confirmation from thought and reason. The growth and knowledge generation from the neuronal maturation of the brain are relational to the cultural transmission of social interaction. Developmental processes consist of the acquisition of knowledge from learning during critical periods. The developing mind is an inferential machine for the generation of knowledge and the maturation of brain circuitry.

The interaction of biological to digital computation entails the knowledge generation of informational content from mental computation into machine computation. The information production from the interaction of biological to digital computation comprises a transformation of evidence from the physical to the digital world. The interaction of digital to digital computation consists of the knowledge extraction of information content from digital resources for the purpose of data analysis and data informatics in the digital world. The interaction of digital to biological computation entails the knowledge extraction of information content from biological datasets into mental computation. The information production from the interaction of digital to biological computation consists of the transformation of evidence from the digital to the physical world.

The functional purpose of machine learning is to understand the regularities and patterns in the structure of the world from the interaction of minds and machines. The learning of biological machines describes mental performance for knowledge generation. The learning of computing machines describes information production for knowledge extraction. The causal power of minds and machines for machine learning demonstrates the malleability of mental content. The causal power of machine learning on mental performance and information production is the correspondence of mental content from truth theories with the structure of the world.

Mind and Machine Learning

Mental inferences arise as detection of the regularities in the structure of the world and its interactivity with the mind. Computational models provide an important approach to understanding the source of mental inference from machine computation. Biological machine computation refers to the physical states and their transitions in the biophysical mechanisms of the nervous system. The computational modeling of neural networks allows for the formal testing of models through data fitting. The formal testing of models in the activation patterns of neural networks aims to identify causal explanations from large datasets.

Machine learning describes mental content as a causal power of realizers of physical property. Mental property as states of explanatory or causal inferences about the structure of the world and its relations supervenes on neural states as physical realizers. The causal power of mental content implies that mental states supervene on physical property as physical events. Biological machine learning refers to the learning from the cooccurrence of neuronal activation and its relation to mental inference. Machine learning as a computational model refers to the production of mental content based on formal testing and data fitting.

Computational approaches provide tools for the formal modeling and prediction of the relation of mental inference to patterns of neural activation. Machine learning algorithms such as pattern recognition allow tools to identify patterns of neural activation that predict categories of mental content based on learning rules. Mental inferences as the byproduct of neural networks entail the flow of information across layers of information processing. Patterns in the activation states of neural networks describe the transformation of information flow from input to output through interactivity. The autonomous brain dynamics of neural networks contribute to the prediction of mental inference.

The computational principles of learning consist of distinct mechanisms for understanding the relations of patterns of activation and mental inferences. Bidirectional neural networks describe a pattern of activation as a mechanism underlying changes in the directionality of patterns of thought and mental content. Inhibitory interactions, such as feedback and feedforward inhibition, enhance selectivity as relations of neural activation and patterns of thought. Deep learning as a learning algorithm refers to the activation patterns of bidirectional networks through multiple hidden layers. Mental inference from deep learning consists of the classification of information based on rules of similarity.

The contextualization of thought and reason arises from learning as explanatory inference. Thought as a flow of information in neural networks undergoes transformation and interpretation from sense data to prediction. Contextual representations of recurrent neural networks perform the change of transformations or interpretations of sense data into output within the hidden layer representations of neural networks. The copying of a hidden layer of network activations into the context layer preserves information and performs learning on internal representations for prediction. The contextual layer of recurrent neural networks demonstrates the malleability of internal representations of the mind. Thought and reason from contextual representations imply a standard of criteria for explanation and causation.

Explanatory inference as thought and reason describes the flow of information from the principles of brain dynamics. Autonomous brain dynamics consist of the flow of information of matter and energy based on local attractor dynamics. Explanatory inference as an optimal state refers to the satisfaction of multiple constraints based on environmental inputs and states of network activation at a lower level of energy. Understanding the causal patterns in the activation states of neural networks contributes to prediction and explanation.

Explanatory inference as an inference to a complete explanation is comprised of the parts and particulars of the total causal structure in the world. The interactivity of parts and particulars in the total causal structure in the world entails a relational correspondence of the data and hypotheses. The amplification of features as activation within a neural network implies a truth correspondence in the mental content from truth theories with the causal structure in the world.

Culture and Machine Learning

Culture as a causal power entails that cultural property as mental property supervenes on physical property as its parts and particulars. Cultural property supervenes on physical property as a physical event. Culture as mental causation is the mental and physical property from the interaction of the mind in the world. Culture as mental causation entails the supervenience of mental property on physical property as a correspondence of mental content with the structure of the world.

TABLE 10.2 Culture as machine learning

	Mind	*Machine*
Model input	Sensation	Dataset
Model algorithm	Learning rules	Data fitting
Model output	Mental inference	Prediction
Purpose	Knowledge generation	Knowledge extraction

The causal power of cultural computation is the knowledge generation from the mental computation of the cultural sphere. Computational models of cultural learning are an approach to the formal testing of the source of cultural inference from cultural machine computation. The cultural model is the formal testing and prediction of the transformation of sensation into mental inference. The cultural model is a formal testing of hypotheses from cultural patterns of activation in neural networks.

Culture as machine learning entails understanding the cultural patterns in the causal structure of the world. The cultural machine learning from the interaction of minds and machines describes the mutual productivity of knowledge generation and knowledge extraction from computation (**Table 10.2**). Cultural patterns in the causal structure of the world inform observation and true belief. Cultural patterns of thought as explanatory inference describe distinct units of explanatory inference as parts of the complete explanation.

The purpose of cultural machine learning algorithms is to understand the regularities and patterns for the prediction of cultural inferences from the brain. Cultural machine learning consists of conceptual and methodological tools for the computational modeling and prediction of cultural inferences from cultural patterns of brain activation. Cultural machine learning algorithms predict cultural inferences from patterns of brain activation and brain dynamics. Bayesian approaches to the cultural brain describe the prediction of cultural inference from encoding in patterns of neural activation and brain connectivity.

Culture in artificial neural networks is the computation modeling of cultural neural networks. Simple approximations of the activation states of neural networks describe a formalization of the computational properties in neural networks and their causal relations. Culture in artificial neural networks consists of the net functions as machine properties.

At the cultural system level, machine learning describes computational approaches to the modeling and prediction of inferences in the cultural sphere. Computational approaches consist of the design of theoretical models for the prediction of cultural inferences. Computational models of culture consist of the formal testing of rule-based learning and data fitting. Cultural machine learning is the knowledge extraction from patterns of information in large datasets. Cultural machine learning algorithms perform data extraction of learning rules as cultural patterns from large datasets.

Cultural programs designed for cultural machine learning perform knowledge extraction. Cultural programs perform data fitting to the cultural model for predictions from large datasets. Cultural programs perform data extraction to reinforce predictions of patterns and regularities from cultural input. Cultural machine learning is the knowledge extraction from cultural datasets based on cultural patterns and regularities in the structure of the world and its interactivity. Culture as the mental content from deep learning is the cultural patterns of activation from bidirectional networks across multiple hidden layers. Cultural content from the deep learning algorithm is mental content from the performance of the transformation of information and its interpretation in multiple hidden layers.

Cultural Machine Learning

Cultural learning is the process of knowledge acquisition in the cultural sphere. The learning of cultural processes requires cultural transmission and knowledge generation from cultural model to learner. The cultural transmission of social knowledge from cultural model to learners relies on the transfer and persistence of informational content across minds. The knowledge generation of learners is the acquisition of mental constructs based on interaction with environmental input. The knowledge acquisition of cultural information for social communication and social interaction contributes to cultural adaptation.

Cultural machine learning is designed to test simple models of hypotheses for prediction of cultural inferences. Cultural machine learning is comprised of the rule-based algorithms for explanation and causation. Cultural content as explanatory inferences is the transformation of input from rule-based algorithms into inferential explanations. The cultural model as a test of hypotheses in cultural neural networks is the prediction of inferences from cultural patterns of neural activation in brain regions and their connectivity. The cultural machine learning algorithms that predict cultural inferences detect cultural patterns that inform observation and true belief of the structure of the world.

Cultural mental computation performs theory confirmation based on cultural thought and reasoning. Cultural models test formal hypotheses of cultural patterns of activation in bidirectional networks with multiple hidden layers. Cultural model nets consist of cultural patterns of neural activation that act as mechanisms that change the directionality of cultural patterns of thought and cultural mental content. Cultural model nets perform the transformation of sense data into representations and interpretations to output. Cultural mental inference from deep learning is comprised of the classification of information content that is rule-based cultural content.

The contextualization of cultural thought and reason in machine learning arises as culture-specific explanatory inference. The contextualization of cultural representations in recurrent neural networks performs the transformations or interpretations of sense data into output within the hidden layer representations. The cultural patterns of network activation as a copy of the hidden layer of the network perform the learning on internal representation for prediction. Cultural thought and reason entail a standard of criteria for explanation and causation.

Cultural machine learning algorithms of the biological computing machine generate knowledge from the cultural transmission of social interaction. Cultural machine learning algorithms perform data fitting to a cultural model to reinforce patterns and regularities from the cultural input. The transfer and persistence of informational content across minds demonstrate the knowledge generation of learners from social interaction with environmental input. Social interaction with environmental input shows the malleability of mental content. The developing mind as a biological computing machine that predicts cultural inferences generates cultural knowledge with the maturation of cultural brain circuitry. The developing mind as a cultural computing machine consists of the neural mechanisms for the acquisition of cultural knowledge and cultural learning.

Cultural programs that perform machine learning perform knowledge extraction of informational content from large datasets. Cultural machine learning algorithms perform model fitting on large datasets and produce output. Cultural machine learning algorithms perform model fitting to datasets to satisfy parameters of cultural programs. The functional role of machine learning from cultural programs is the demonstration of the biological plausibility of cultural plasticity in the biological organism. The knowledge extraction from cultural machine computation is the identification of causal roles in the patterns and regularities in the structure of the world and its interactivity.

Cultural adaptation refers to the knowledge acquisition and cultural learning of the functional processes of cultural competence and cultural communication based on cultural contact. The maintenance of positive attitudes of heritage and host cultures contributes to cultural competence and cultural communication. Cultural adaptation implies the functional equivalence of mental content across heritage and host cultures. The maintenance of heritage and host culture and the promotion of social relations of ethnic and cultural groups strengthen the processes of cultural adaptation.

Conclusion

In philosophy of mind, the role of change in mental thought and mental content arises from the understanding of computational principles of interactivity. Machine learning entails that change in mental thought and mental content supervenes on physical property. Change in the directionality of mental thought and mental content implies a change in physical property across levels of analysis from the principles of computation. In philosophy of science, machine learning plays a functional role in theory confirmation and evidence. The functional role of minds and machines is the performance and production of explanatory inference to the complete explanation. Machine learning at the cultural system level consists of the modeling and prediction of cultural inferences in the cultural and public sphere. Computational models of culture perform the formal testing and prediction of cultural patterns based on rule-based learning and data fitting from large datasets. Machine learning contributes to cultural change and cultural adaptation.

Cultural machine learning entails the knowledge extraction from cultural data for the simulation and construction of cultural production. Cultural machine learning tests simple models for the prediction of cultural inferences. Cultural models test hypotheses of cultural neural networks from cultural patterns of neural activation in brain regions and their connectivity. Cultural inferences from cultural machine learning algorithms detect cultural patterns of information observation and true belief of the structure of the world.

References

Churchland, P.S. & Sejnowski, T.J. (1992). *The computational brain*. Cambridge, MA: MIT Press.

Hebb, D.O. (1949). *Organization of behavior*. New York: Wiley.

11

INTELLIGENCE

Introduction

Philosophical notions of mind are fundamental to explorations into the content and boundaries of intelligence. The philosophical inquiry into the nature of intelligence entails considerations of mentality, the malleability of intelligence, and the contextualized patterns of thought and reason. Early conceptualization of intelligence sought to determine generalizations of the human mind. Contemporary approaches seek to explain intelligence in terms of scientific theories of thought and reason across levels of analysis. In the normal science sense, scholarly interest in intelligence is an endeavor to advance a scientific approach to mental life. In the broadest sense, explorations of intelligence complement a humanistic understanding of the nature of the mind.

Intelligence

Scientific approaches to the study of intelligence seek to describe the variety of mental capacities that comprise thought and reason. Mental capacities of thought and reason facilitate the attainment of goals and the sense of satisfaction in explanations of natural phenomena. Intelligence is the thought and reason that informs the plans and actions of individuals and groups. Intelligence consists of multilevel forms of task performance and information production. Intelligence informs the capabilities of decision making and strategic defense.

At the individual level, intelligence refers to the mental capacities of individuals for thought and reasoning. Intelligence is conceptualized as the mental computation required for task performance. The intelligence of individuals consists of the performance on a range of mental tasks, from cognition and emotion to social reasoning and decision making. Cognitive intelligence refers to the capacity of individuals for performance of cognitive tasks (Nisbett, 2010). Emotional intelligence concerns the capacity for emotional knowledge and emotional control (Brackett, Rivers, Salovey, 2011). Social intelligence is related to the capacity for social knowledge and social regulation. Social intelligence is related to the complexity of social networks and the social brain hypothesis, or the functional specialization of the brain for social processes (Dunbar & Schultz, 2007).

At the group level, group intelligence refers to the group performance on a range of tasks. The intelligence of groups refers to the performance of groups on a wide range of tasks. Collective intelligence refers to the group performance on a wide range of tasks through collaboration and cooperation (Woolley, Chabris, Pentland, Hashmi, Malone, 2010). Collective intelligence is related to the capability in the performance of groups that is independent of the capacity of individuals. Collective intelligence is related to group factors such as social sensitivity, conversational turn taking, and gender participation.

At the national level, national intelligence refers to the strategic capability of nations to provide information for the purpose of policy making and national security (Johnson, 2010). National intelligence is related to the strategic capability to provide information for policy-informed decision making in the political, economic, military, or diplomatic sphere. National intelligence is related to the production of informational content, the process of analysis and dissemination of information for policy makers, missions of counterintelligence and covert action, and operations that include clusters of people and organizations for the performance of national intelligence missions. Strategies of national intelligence seek to gather information that is accurate, timely, and reliable for purposes of policy making and governance.

At the cultural level, cultural intelligence comprises the performance of cultural groups on a wide range of tasks. Cultural intelligence refers to the cultural patterns of thought that contribute to cultural participation and cultural life. Cultural intelligence consists of the cultural patterns of thought that relate to tradition, customs, practices, values, and beliefs of groups of people. Cultural intelligence is related to the performance of cultural tasks of ethnocultural groups that promote group goals of societal equality, social inclusion, and cultural diversity. Cultural intelligence refers to the performance on cultural tasks that affects the social relations of ethnocultural groups. Cultural intelligence informs the strategic capabilities of nations for peace and security. Cultural intelligence includes the strategies and programs for cultural development and the advancement of policy-informed decision making in the cultural sphere. Strategies of cultural intelligence seek to promote international cooperation in the cultural sphere.

Multiple Intelligence

Multiple intelligence refers to the notion that the mind performs mental tasks across multiple domains. The mind demonstrates mental capacities for thought and reasoning across multiple domains. Multiple intelligence places emphasis on a broad viewpoint of mental thought and reasoning. Multiple intelligence acknowledges the multitude of patterns of mental thought that comprise intelligence, rather than a generalized notion of intelligence (Gardner, 2011).

The theory of multiple intelligence supposes that there exist multiple domains of intelligence. The multiple intelligence domains consist of functionally specialized operations that are discrete and autonomous. The multiple domains of intelligence are physical, instantiated in the mind as biological adaptations. Across evolutionary history, the functional architecture of the human mind and brain demonstrates multiple domains of specialized information-processing mechanisms that operate independently. The biological adaptation of the human mind and brain underscores the relevance of functional specialization to computational components of intelligence.

Multiple domains of intelligence as physically instantiated in the human brain suggest multiple information-processing mechanisms for thought and reasoning. Information-processing mechanisms for thought and reasoning demonstrate several aspects that contribute to the multiple viewpoint of intelligence (O'Reilly & Munakata, 2000). Multiple intelligence domains highlight the multilevel nature of the structural and functional organization of the nervous system. The functional specialization of biological machines implies that thought and reasoning arise through information processing that occurs in parallel or simultaneously. Parallel distributed processing refers to the parallel operation of the cognitive components in biological machines (McClelland & Rumelhart, 1986). The information processing of cognitive functions operates in parallel across individual processing mechanisms located throughout layers of cortical organization. The parallelism of cognitive functions implies that multiple information-processing mechanisms, rather than a generalized form of processing, constitute intelligence. The neurobiological basis of cognition supports the notion of parallelism and the biological plausibility of multiple intelligence.

Multiple intelligence considers the contextual nature of thought and reasoning. The notion of multiple intelligence assumes the capacity of individuals to demonstrate competency in performance across a range of intelligence domains. The information-processing mechanisms of cognition show interactivity or bidirectional connectivity in the flow of information across neural nets. The cognitive properties of interactivity allow for top-down and bottom-up influences on the information flow of processing mechanisms. The information flow in representational and interpretive layers of neural nets is malleable to the directionality of information processing and the contextual influences of interactivity. The contextual influence of interactivity on information-processing mechanisms describes how the directionality of information flow demonstrates the malleability of intelligence.

Intelligence as Computation

Early philosophical notions of intelligence as computation arise from the premises of machine functionalism. The proposal of machine computation, or Turing machines, introduces considerations brought on by the possibility of functional equivalence in mentality and intelligence. The Turing test asserts that computing machines can be programmed to perform at a level of functional equivalence to the human mind (Turing, 1950). The mere suggestion of equal performance in humans and machines unfolds a multitude of questions regarding the meaning of intelligence in mental life.

The mind as a computing machine demonstrates the capacity for mentality and intelligence (**Table 11.1**). The mentality and intelligence of the mind rely on the capacity to perform mental tasks at the level of computation. Mentality and intelligence suggest that mental computation is sufficient for task performance, and that the performance of mental tasks is a fundamental standard of computation. The equal performance on mental tasks of humans and machines implies equivalence in mentality.

The causal power of minds and machines to demonstrate intelligence or the capacity for performance of thought and reasoning gives rise to several considerations. Functional equivalence in intelligence of minds and machines suggests a commonality in the functional roles of humans and computers. The production of information content from mental and machine computation that is functionally equivalent is consistent with the criterion of performance that comprises intelligence. The natural intelligence of the human mind is the source of the production of informational content through the computational components of the brain. The artificial intelligence of the computer is the source of the production of information content through the computational components of the computing machine.

The functional equivalence of mental and machine computation suggests that natural and artificial forms of intelligence share functional properties at the level of algorithm that are independent of its physical implementation. The algorithmic

TABLE 11.1 Intelligence as computation

	Mind	*Machine*
Computational	Problem solving	Programming
Algorithm	Mental function	Formal procedure
Physical implementation	Biological	Computer

level describes the formal procedure for the production of accurate output given an input. The independence of algorithm from its physical implementation implies that the production of information content is malleable to multiple physical realizers.

Other properties of mental and machine computation pertain to the trade-offs of intelligence in natural and artificial forms. Intelligence as the flow of information from thought and reason contains several properties, including accuracy, reliability, and efficiency. The truth value of thought and reason is a property that affects the informational content of intelligence; the accuracy and reliability of informational content are necessary for the capability of intelligence to inform decision making and plans of action. The efficiency of thought and reason affects the causal power of intelligence. Intelligence as information production that occurs automatically demonstrates a causal role in the complex network of mental life.

The incommensurability of mental and machine computation arises from considerations of the meaning of intelligence in mental life. The intelligence of mental computation is an emergent property of biological machines for the purpose of adaptation. The intelligence of machine computation is a functional property of computing machines for the purpose of complexity. The functional equivalence in computational performance of minds and machines is not necessarily a commonality in functional purpose. This distinction in the functional purpose of computation implies that the intentionality of design matters and is inherently different in minds and machines.

The meaning of intelligence in mental life broadens into considerations of consciousness. Philosophical notions of consciousness in mental life place emphasis on the importance of experience as knowledge. The experience of mental life has a qualia or feeling of what it is like that is a part of the truth of mentality. Consciousness as a property of natural forms of intelligence relates to the knowledge of self-awareness. The consciousness state of mentality consists of the stream of subjective experience that comprises mental life. The mental property of consciousness that is a part of natural intelligence and mental computation supervenes on physical property. The supervenience of consciousness on physical property implies that states of consciousness have computational components and the experience of consciousness is fundamental to intelligence and mental life.

Intelligence and Computation

At the level of computation, intelligence refers to the functional components of problem solving and mental task performance. In mental computation, intelligence consists of the set of mental tasks for problem solving. In machine computation, intelligence refers to the component parts that are important for task performance. Across domains, intelligence at the computational level consists of the hierarchical structure of tasks and task components for goal performance.

Mental performance that requires cognition refers to the problem solving of cognitive functions. Cognitive functioning is related to the capacity to solve problems that require knowledge and reasoning skills in perceptual, spatial, linguistic, or numerical domains, among others. Cognitive functioning consists of the set of mental tasks that demonstrate capacities for reasoning, memory, rule use, inhibition, and control. The use of cognition in mental task performance is an example of a type of intelligence. The performance of cognitive mental tasks implies the presence of mentality and intelligence.

Emotional intelligence refers to the mental tasks related to capacity for emotional knowledge and emotional control. Understanding the emotional states of people relies on emotional knowledge and the control of emotion. Emotional knowledge is related to the perception, recognition, and expression of emotion in nonverbal communication. Emotional knowledge of one's self requires emotional awareness, or the awareness of the emotional states of the self, from introspection and subjective experience. Knowledge of the emotions of others requires the social reasoning of emotional mental states from verbal and nonverbal cues in social communication. Emotional control is an important component of emotional intelligence. Emotional control refers to the appropriate use of emotion in social and cultural contexts. Emotional control relies on the capacity for the regulation of emotion. The presence of emotional knowledge and control in social and cultural contexts is fundamental to individual and group performance.

Social intelligence refers to the performance on mental tasks that rely on the capacity for social knowledge and social regulation. Social intelligence is related to the capacity to understand the mental states of other people, including intentions, beliefs, and goals and other internal states. Social intelligence requires awareness of the self and others and social reasoning from verbal and nonverbal communication. Social knowledge of self and others consists of social trait attributes and understanding of the contextual nature of social attribution. Social reasoning comprises the processes of inference related to perspective taking, theory of mind, and empathy. Social reasoning is guided by inferential thought from social knowledge and social experience. Social regulation refers to the capacity to regulate social thought and action and to conform to social norms. The presence of social knowledge and social regulation in cultural contexts is important to the performance of individual and group goals.

Collective intelligence is related to the performance of groups on a range of tasks. Collective intelligence consists of the ability of the group to perform goals based on the group. Collective intelligence implies that the functional components for problem solving are based on group dynamics, or the interactive patterns of thought in the group, rather than individual capacities per se. Collective intelligence is related to social sensitivity, conversational turn taking, and gender participation.

National intelligence is the production of information for decision making in the political sphere. National intelligence refers to the capability of individuals and groups to perform strategic intelligence missions and operations. National intelligence missions consist of strategic operations in counterintelligence and covert action. The production of information content that informs decision making in the political sphere relies on strategies of national intelligence.

Cultural intelligence refers to the cultural mental tasks that are related to traditions, customs, values, practices, and beliefs of groups of people. Cultural intelligence consists of the mental capacities that comprise cultural patterns of thought. Cultural intelligence is the performance of cultural tasks that are important to group goals of societal equality, social inclusion, and cultural diversity.

Across domains of intelligence, the computational level consists of an analysis of the set of tasks required for the performance of goals. The computational level details the decomposition of functional tasks into component parts across domains. The computational level comprises a problem analysis that is required for task performance. The computational level is independent of other levels of analysis and generates a hierarchical structure of component parts for task performance.

Intelligence is deterministic to the level of algorithm in mental computation. At the level of algorithm, intelligence is the functional performance of accurate output given specific input. The capacity to produce an accurate output given a specific input defines the performance standard of intelligence. The production of accurate output is necessary and sufficient for the assumption of reliability in intelligence.

Mental computation can produce accurate output from multiple algorithms. The production of accurate output in mental computation from multiple algorithms assumes multiple formal procedures for the production of specific output. Multiple algorithms suggest a set or range of strategies for equivalence in problem solving. Multiple algorithms imply that problem solving is not fixed, limited, or immutable to a specific procedure or strategy per se. Multiple algorithms demonstrate the malleability of the algorithmic level to a wide variation in patterns of thought and reasoning. Multiple algorithms assume the importance of contextual factors in patterns of thought and reasoning.

At the level of physical implementation, intelligence is instantiated in multiple physical realizers. The algorithmic level is independent of the level of physical implementation. That is, the mental capacity to perform specific functional tasks is independent of the physical realization of its computation. Likewise, intelligence as the mental capacity to perform mental tasks is independent of physical implementation. The mental computation of biological machines is a form of natural intelligence. The machine computation of computing machines is a form of artificial intelligence. From minds to computers, computing machines demonstrate the functional capability of intelligence.

Intelligence in Culture

Intelligence plays a foundational role in the production of culture. Cultural values, beliefs, and practices refer to the patterns of thought that define cultural groups.

Cultural patterns of thought demonstrate the role of intelligence in cultural mental life. Cultural processes of the mind and brain generate the mental content for the creation and maintenance of culture. Cultural processes refer to the creation and maintenance of mental content for cultural adaptation. The cultural computation of machines contributes to the production of informational content as cultural property or the property of the cultural system level.

Cultural property refers to the physical and mental properties that comprise shared meaning systems that define groups of people. Cultural property includes the property of values, practices, and beliefs of groups and individuals at the cultural system level. Cultural capital refers to the valuation or worth of cultural property. The protection of cultural capital is the purpose of cultural security.

Cultural intelligence as patterns of thought refers to cultural processes and their underlying neurobiological basis. Mental computation is foundational to the structure and dynamics of cultural and evolutionary processes. The functional operations of mental computation comprise the set of mental capacities that contribute to cultural evolutionary processes. Cultural adaptations that demonstrate a functional specialization are considered a part of cultural intelligence. Culture guides the functional architecture of the computational components of intelligence. The natural intelligence of the human mind and its biological basis is a fundamental component of culture life.

Cultural processes consist of cultural dimensions, orientations, and systems that define cultural patterns of thought. Cultural participation in the traditions and practices that define groups of people relies on the processes of cultural intelligence. The development of cultural intelligence pertains to the building of awareness of the importance of commitment and belonging in the cultural community. Cultural advancement entails the strategic use of cultural intelligence to inform the development and implementation of plans of action and decision-making strategies for policy making and governance.

Culture is fundamental to the artificial intelligence of computers. Early machine computation is built on the demonstration of the capabilities of artificial intelligence at the cultural system level. The earliest computers demonstrated a functional role for artificial intelligence in informational production at the level of cultural systems. The standards of performance in artificial intelligence are defined in the development and implementation of multilevel strategic capabilities in cultural systems. Cultural systems rely on the strategic capabilities of computation for the informational production of resources for system-wide performance in the public sphere.

The cultural properties of computation underscore the functional capability of natural and artificial intelligence for the cultural acquisition of knowledge and the cultural transmission of patterns of thought. The cultural computation of natural cultural intelligence refers to the information-processing mechanisms of the nervous system for cultural knowledge acquisition and cultural learning. Natural cultural intelligence facilitates processes of cultural inheritance or the transmission of cultural processes across related and nonrelated individuals. Cultural intelligence refers to the encoding and decoding of cultural information in the mechanisms of mental computation.

Cultural artificial intelligence refers to the computational models and algorithms of machines for the simulation and construction of cultural production. Cultural artificial intelligence consists of the computational components of machine computation that contribute to the simulation and construction of information production at the cultural system level. Programs that demonstrate cultural artificial intelligence include the programming code, parameters and environment for the design and automation of cultural routines and tasks. Computer programs for the production and automation of cultural routines demonstrate the performance capabilities of cultural computation from computing machines. The production of cultural information from machine computation is an automated form of cultural property.

Fundamental standards of functional equivalence in mental and machine computation are operationalized at the cultural system level. Natural forms of cultural intelligence seek to generate and maintain cultural knowledge systems of the social and physical world. Natural forms of cultural intelligence protect cultural capital through cultural processes of niche construction, including cultural transmission and cultural inheritance. Artificial forms of cultural intelligence aim to construct cultural property and to protect cultural capital through computational processes. Cultural intelligence in artificial form facilitates the design and production of cultural property and the protection of cultural capital through its reconstruction and replication.

Multilayer strategic cultural intelligence provides the functions of defense in cultural security. Strategic cultural intelligence consists of the multilayer system-wide defense for purposes of cultural security. Strategic cultural intelligence refers to cultural strategies for the protection of social and physical property at the cultural system level. Cultural strategies consist of the programs and standards that inform the social function of defense capabilities. Cultural strategies are responsive to the severe and pervasive threats that affect security and provide the necessary resources for protection and recovery. Cultural strategies are effective for the preparation and prevention of security threats. Strategic cultural security controls threats and protects core interests through the development and implementation of programs at the cultural system level.

Cultural development refers to the empowerment of individuals and groups for the cultural production of resources and capabilities. Cultural development pertains to the access to education and information for the development of social patterns and plans of action. Programs that build on the production of cultural resources enhance capabilities for decision making and governance. Cultural development programs that empower instill a sense of belonging and commitment of individuals to their cultural community. Programs of cultural development are an investment in the cultural capital of the community. Cultural development programs contribute to the strategies of prevention and intervention in cultural security.

Participation in cultural life is one of the most fundamental of human rights. Cultural life is the source of cultural development of individuals, societies, and nations. The development of cultural activities and the maintenance of cultural traditions promote capabilities of protection and empowerment in the cultural sphere. Cultural patterns of thought guide societal perception towards the fulfillment of human rights and human development. Cultural advancement is a societal achievement towards the fulfillment of cultural and human rights.

Culture and Intelligence

Cultural constructs define the structure and function of informational content and its causal power in the cultural system level. Cultural constructs consist of the multilevel organization of the flow of information in matter and energy of cultural systems. Changes in cultural constructs are reflected in the information flow as physical state transitions that illustrate the functional relations of mental and physical property in the cultural system. Cultural constructs describe the functional properties of information content and the cultural causal networks that contribute to cultural production. Cultural constructs consist of the standards and programs for the production of informational content at the cultural system level of nations, institutions, and individuals.

Cultural constructs describe the causal functional role of mental and machine computation at the cultural level. Mental and machine computation describes the production of informational content of cultural systems. The cultural computation of minds and machines entails the functioning of independent and interactional properties of computational components for cultural production. Cultural constructs depict the content, function, and use of information content at the cultural level.

Cultural property as the mental and physical property of cultural systems consists of the informational content that comprises intelligence in the environment (**Table 11.2**). Cultural property relates to the mental and physical property of culture as an organized system with parts. The cultural property of symbols, heroes, and rituals consists of the mental and physical property produced through

TABLE 11.2 Cultural computation

	Mental	*Physical*	*Machine*
Cultural property	Mental property	Physical property	Machine property
Cultural capital	Mental capital	Physical capital	Machine capital
Cultural security	Human security	Homeland security	Machine security

the construction of the social and physical environment. The cultural property of values, practices, and beliefs refers to the relations of mental and physical property of organisms that comprise intelligence at the cultural level. The mental and physical property of culture entails the informational content from mental computation. The machine property of culture is comprised of the information production from machine computation.

Cultural capital is the valuation and worth of cultural property. Cultural capital refers to the valuation and worth of the mental and physical property based on causal functional roles of the cultural system. Cultural capital is the valuation and worth of mental and physical capital from mental computation at the cultural level and the machine capital from machine computation of the cultural system. Cultural capital is affected by the causal power of cultural property. Cultural capital pertains to the cultural patterns of functional performance in mental and physical property. Cultural capital relates to the causal power of cultural property for cultural production.

The protection of cultural capital is a functional role of strategic cultural intelligence. Cultural capital entails the valuation and worth of cultural property from an initial truth state. Strategic cultural intelligence protects cultural capital through the automation and control of cultural production. Cultural strategies provide protection through the multilevel functions of cultural processes. Cultural strategies seek the most harmonious states for cultural property in the cultural causal network. Strategic cultural intelligence automatically recognizes the multiple constraints of the environment imposed on the causal network of cultural property.

Cultural capital is an indicator of the cultural resources for cultural development and cultural security. Cultural capital is related to the level of human development and human security. The valuation and worth of cultural capital are affected by the level of human development and protection from threats in human security. Cultural capital contributes to the capabilities of protection and the security of states and people. High levels of cultural capital (e.g., high cultural protections, low cultural conflict) demonstrate the elements of societal arrangement for cultural development and cultural security.

Cultural security consists of the use of strategic cultural intelligence for the purposes of defense. Cultural security includes the capabilities for the development and implementation of cultural programs and standards for cultural production. Strategic cultural intelligence informs the defense capabilities of programs and plans of action at the cultural level. Strategic cultural intelligence informs the capabilities for decision making and governance that protect cultural security.

Cultural Intelligence

Cultural intelligence affects the capacities of individuals and the capabilities of groups to perform functions of protection at the cultural level. Cultural intelligence refers to the cultural patterns of thought that contribute to the societal perception and traditions of groups of people. Strategic cultural intelligence provides informational content for the decision making of individuals and groups across levels.

At the national level, strategic cultural intelligence informs the development and implementation of plans of action that comprise national policies. Cultural intelligence consists of the information content that contributes to evidence-based decision making of policy makers and governance. Cultural intelligence informs the intention and commitment of stakeholders for action in the political and cultural sphere. Strategic cultural intelligence is foundational to the priority setting and attainment of goals of ethnocultural groups.

At the institutional level, strategic cultural intelligence contributes to the development and implementation of programs and standards that inform the cultural level. Cultural intelligence is a source of rationale that supports the identification and priority setting of group goals, the establishment and reform of programs, the marshaling and strengthening of resources, and the use of evidence for decision making. Institutional infrastructure demonstrates policy implementation and the function of management arrangements as implementation resources for institutional initiatives. Institutional programs provide the organizational structure for the delivery of services and resources based on institutional intent and commitment. The use of strategic cultural intelligence for the implementation of institutional commitment requires the monitoring and evaluation of strategies and outcomes at the cultural level.

Strategic cultural intelligence informs the decision making and policy making that protect diversity, equity, and inclusion at the institutional level. Cultural intelligence contributes information and evidence-based strategies for the protection of institutional policies and practices of cultural diversity, equity, and inclusion. Institutional policies and procedures that protect cultural diversity, equity, and inclusion demonstrate the implementation of cultural programs and initiatives at the institutional level. Institutional initiatives that promote the development of strategies for cultural diversity, equity, and inclusion build and strengthen the institutional resources for policy implementation. Building awareness of the issues of cultural diversity, equity, and inclusion informs the decision making and actions of institutions.

At the individual level, cultural intelligence is informative of strategies for multicultural ideology. Multicultural ideology reflects an appreciation of the value of cultural diversity in plural society. Multicultural ideology contributes to cultural strategies that promote societal practices of cultural diversity. Cultural strategies that support cultural diversity include the promotion of social relations among ethnocultural groups through positive attitudes and an appreciation of cultural heritage.

Conclusion

Fundamental notions of intelligence build on the notion of mental computation. Intelligence is a form of mental computation of individuals and groups. The mentality that comprises intelligence relates to the neural mechanisms of thought and reason. Specialized functional mechanisms of the brain perform specific mental functions based on response to environmental conditions.

The intelligence of groups refers to the performance of the group on a range of tasks. The intelligence of groups refers to the dynamic interaction of collaboration and cooperation that contributes to problem solving. The intelligence of groups arises from the flow of information across individuals during social interaction. The mental computation of groups may be characterized in computation with dynamic causal modeling of mental computation in social interaction.

The intelligence from mental computation of individuals refers to the patterns of thought and reason that guide the performance of tasks. Patterns of thought and reason generate informational content that contributes to the performance of functional tasks. The intelligence of individuals demonstrates the causal power of thought and reason in mental life.

References

Brackett, M.A., Rivers, S.E., Salovey, P. (2011). Emotional intelligence: implications for personal, social, academic and workplace success. *Social and Personality Psychology Compass, 5(1)*, 88–103.

Dunbar, R.I. & Shultz, S. (2007). Evolution in the social brain. *Science, 317(5843)*, 1344–1347.

Gardner, H. (2011). *Frames of mind: The theory of multiple intelligences.* New York: Basic Books.

Johnson, L.K. (2010). *The Oxford handbook of national security intelligence.* New York: Oxford University Press.

McClelland, J.L. & Rumelhart, D.E. (1986). A distributed model of human learning and memory. In McClelland, J.L., Rumelhart, D.E., & PDP Research Group (Eds.). *Parallel distributed processing. Volume 2: Psychological and biological models* (pp. 170–215). Cambridge: MIT Press.

Nisbett, R.E. (2010). *Intelligence and how to get it: Why schools and cultures count.* New York: W.W. Norton.

O'Reilly, R.C. & Munakata, Y. (2000). *Computational explorations in cognitive neuroscience: Understanding the mind by simulating the brain.* Cambridge, MA: MIT Press.

Turing, A. (1950). Computing machinery and intelligence. *Mind, 59*, 433–460.

Woolley, A.W., Chabris, C.F., Pentland, A., Hashmi, N., Malone, T.W. (2010). Evidence for a collective intelligence factor in the performance of human groups. *Science, 330(6004)*, 686–688.

12

VIRTUAL REALISM

Introduction

Virtual realism is a philosophical inquiry into the foundational notions of the nature of mental life and the role of mental life in the natural world. Mental life is the mental content of the mind and its causal role in the natural world. Mental life is the generation of the mental events that act as a causal power in the physical world. The complexity of mental events as a causal system illustrates the complexity of mental events as physical events in the physical world. Similarly, the notion of the complexity of mental events as a causal system is preserved in the notion of the complexity of mental events as virtual events in the virtual world. Virtual realism is a reflection into the content of the mind and its causal role in the mental life across possible worlds.

Virtual realism refers to the philosophical positions of scientific realism. Scientific realism posits that reality exists in the structure of the world independently of the mind, except for mental content that exists as parts of reality with a causal role in the world (Godfrey-Smith, 2003). Virtual realism as a position of scientific realism refers to the structure in the world that is independent of societal perception, to the extent that mental content is a reality in the world. Virtual realism as a part of scientific realism is a product of scientific theory at the level of computation.

Virtual realism contributes to the position that the scientific theory of computation describes a real dimension of the structure of the world. Similar to other philosophical positions of scientific realism, virtual realism functions to provide accurate representations of the structure of the world through the scientific theory of computation. From the physical to virtual world, the notion of reality is consistent across possible worlds.

Virtual realism challenges traditional notions of mental content in scientific realism. Traditional notions of scientific realism define mental content as consistent with the thought and language of the mind. Virtual realism broadens the notion of mental life in the natural world. Virtual realism posits that the reality of the structure of the world exists independently of the mind, except to the extent that mental content is realized in the world with causal power. Mental content with a physical implementation constitutes a part of the reality of the world. Mental content existent in multiple physical realizers entails a real part of the structure of the world.

Because the level of computation defines mental content as having a physical implementation in the world, mental computation or mental content at the level of computation is a level of analysis that is consistent with scientific realism. Virtual realism defines mental content as the mental property from biological to quantum computing machines and the perspectives of their interaction. Virtual realism elaborates on the reality of mental content as consistent with the production of mental content from simple computing to supercomputing machines. Virtual realism assumes that the mental content of computing machines connotes parts of reality with a causal role in the world.

Virtual realism implies that mental computation corresponds to a reality in the structure of the world. Mental computation connotes a reality with a causal role in the structure of the world. Mental computation is a progression in scientific realism to the extent that the description of the structure of world is accurate based on the theory of computational science. Virtual realism refers to the maintenance of mental content about the structure of the world through computational machinery. Virtual realism pertains to the augmentation of the reality of the structure of the world through interaction with computing machines.

Virtual realism strengthens the notion that the knowledge of reality is readily accessible at the computational level. Virtual realism describes the knowledge of the reality that is accessible to a range of computing machines. Virtual realism refers to the knowledge of the reality of virtual worlds that is accessible to the computing mind through interaction with the world. Virtual realism describes the knowledge of reality that is accessible in the world through the supervenience of computing machines.

Virtual realism acts as a continuum of mental content in the structure of the physical world. The truth value of mental representation in the structure of the virtual world is consistent with a computational scientific theory with relational correspondence to the structure of the physical world. The truth value of mental representation in the virtual world shows relational correspondence to the truth value of mental representation in the physical world. Virtual realism provides a mental language of meaning and truth reference for digital representations in the virtual world to physical representations in the physical world. Virtual realism is an achievement in the scientific goal to describe what the world is like in scientific terminology at the computational level.

Virtual Realism

Virtual realism refers to the reality in the structure of the world that is from the mental content of computation. Virtual realism as a form of scientific realism assumes a causal role of mental computation in the structure of the world. Virtual realism pertains to the truth correspondence of scientific theory at the level of computation with the structure of the world. To the extent that virtual realism is an accurate description of the structure of the world, the goal of scientific theory at the level of computation is an achievement.

The mind as instantiated in a physical realization connotes the notion of a set of mappings of mental and physical states. The mapping of mental and physical states can be characterized along distinct criteria of functionality. The level of physical implementation of mental states describes one of the primary levels of analysis in computation. The naturalistic perspective of mental states as instantiated at the neurobiological level refers to the computation of the nervous system. Within the nervous system, the functional mapping of mental and physical states occurs as a set of relations of mental and neural states as psychoneural identities.

The contemporary perspective of mental states as instantiated at the computational level refers to the computation of the computing machine. The performance of mental computation of simple to complex machines demonstrates the capability of devices for the production of mental states. The multiple physical realization of mental states in the physical world refers to the capability of a range of machines for the understanding of mental states (**Table 12.1**).

The physical realization of mental states in the virtual world refers to the computational performance of virtual machines. The virtual realization of mental states describes the social capability of virtual machines for social interaction in the virtual world. The multiple virtual realizers of mental states depict the social capability of a range of virtual machines for social interaction in the virtual environment. The technology of machines for computational performance in the virtual world demonstrates a unique capability of a range of machines for the production of reality.

Given that virtual reality is a technological progression in scientific realism that demonstrates the capability for the production of reality, virtual realism presents

TABLE 12.1 Virtual realism and possible worlds

Possible world	Physical	Digital	Virtual
World	Real world	Digital world	Virtual world
Space	Physical space	Digital space	Virtual space
Time	Real time	Digital time	Virtual time
Experience	Real-world experience	Digital experience	Virtual experience
Reality	Physical reality	Digital reality	Virtual reality

novel positions in scientific realism. Virtual realism is a refinement in the notion of reality in the structure of the world, to detail the particular reality that is the production of computation. Virtual realism consists of the production of reality from the mental content at the level of computation. Mental computation that entails the physical realization of mental content in the structure of the world supports the traditional view of scientific realism. Virtual realism describes physical reality in the world that is a production from mental content and refers to the reality in the world from mental content. Philosophical positions of virtual realism further encompass the assumption of the production of reality in the world from computation that is independent of the mind. Because machine computation produces reality in the world independently of the mind, scientific realism entails machine computation. Because the computation of the mind in a physical realizer is a source of the production of reality, scientific realism entails mental computation. Thus, virtual realism presents novel positions of the knowledge of reality in the structure of the world from computation.

Virtual realism refers to the production of reality through the computational performance of machines. The capability of virtual machines for performance of mental computation in the world illustrates the virtual realization of mental states as one of the physical realizers of the mind. The virtual realism of the mind consists of the virtual realization of mental property, including the mental content and mental causation of mental states. The physical realization of virtual mental states demonstrates the complexity of the causal network of mental life in the world.

Virtual mental property consists of mental states with the content and causal roles of mental property in the world. Virtual mental states refer to the mental content that is physically realized in the virtual world. Virtual mental content describes the mental content that comprises the digital representation of the world. Virtual mental content consists of the set of mental states that demonstrate a physical realization in the digital culture of the world. The mental content of virtual mental property refers to the performance of mental computation of machines in the world. The mental content of virtual mental property is characteristic of the complex causal network of mental states that comprises mental life.

Virtual mental causation connotes the causal power of mental content in the virtual world. The causal role of mental content consists of the functional roles for computational performance in the virtual world. The virtual realization of mental causation in the virtual world refers to the functional causal role of virtual mental content in the virtual world. The causal role of virtual mental content may extend beyond the virtual world into other physical worlds. The causal role of virtual mental content consists of the functional input–output generated within virtual mental property into the causal network of mental life.

The mental causation of mental content in the virtual world demonstrates the capability for performance of multiple functionality. The mental content in the complex causal network of mental content demonstrates the functional performance and capability of security and protection of mental life in the virtual world.

The complex causal network of mental content in the virtual world demonstrates the complexity of an adaptive system comprised of mental life.

Mental content in the virtual world is comprised of distinct social representations. The social environment of the virtual world is regulated by distinct kinds of social representation. An avatar refers to the social representation of mental content of human life in a virtual realization controlled by a human. Avatars comprise digital representations of mental content from the human mind. An agent refers to the mental content of human life in a virtual realization controlled by a computer. Agents consist of digital representations of mental content from the computing machine. The social environment of avatars and agents comprises the main types of characters in social interaction with core properties of the virtual world.

The social environment of the virtual world consists of virtual characters. Virtual characters in the virtual social environment may consist of digital social representations with physical characteristics of social beings. The social representation of virtual characters comprised of digital representations of physical characteristics is important for the performance of virtual social behavior in the virtual world. Virtual characters are a main representational unit of performance in virtual social reality.

Mental life demonstrates the performance capability of complex adaptive systems in the virtual world. Complex adaptive systems share core properties that demonstrate the capability for malleability and stability of a living system (Minas, 2014). Mental life in the virtual world is consistent with the properties of complex adaptive systems. The mental content and mental causation of virtual mental life are consistent with the computational performance of a complex adaptive system.

The virtual world as a kind of living system consists of core properties. The notion of virtual life refers to the production of social characters and their physical realizations in the world. Virtual life entails the digital representation and simulation of cultural models in virtual environments. Virtual life includes the multiple replicability of cultural models. Virtual life as avatars consists of social characters as cultural models that can perform virtual scenes. Virtual life as agents consists of social characters that perform virtual behaviors consistent with the functional performance of the cultural model. The multiple replicability of virtual life is a distinct property of virtual social environments in the virtual world.

The virtual world as a life system includes the mental life that is a property of the multiple replicability of virtual life. One functionality of the multiple replicability in virtual life is to ensure the digital performance of cultural models. The digital representation and simulation of cultural models ensure the performance of the functional tasks of the cultural models in the virtual social environment. For functional performance in the virtual environment, virtual life replicability is a computational approach to digital representation and performance of cultural models in the virtual world. The representation and simulation of cultural models require computational modeling of the mental life and other properties of the life system of the cultural model.

Virtual realism as a production of virtual property is a kind of causal property. Virtual property plays a causal functional role in the virtual world. Virtual reality or the realization of virtual property in the physical world is beneficial for the design of technology and its application that is consistent with optimal standards. Virtual reality technology facilitates the development and implementation of standards for the design and criteria of machine performance. The application of virtual reality technology in practice and policy is an example of its performance capability across system levels.

The core properties of mental life in the virtual world imply that virtual life as a living system consists of properties with a fundamental structure. Virtual life and the mental content of virtual life are productions of computation. Virtual life demonstrates the functional capability of technology for the representation and simulation of living systems. The mental content of virtual life represents the production of mental states for living systems. The production of mental content in virtual social environments contributes to the capability and performance of living systems. The production of mental content that is interchangeable in matter and energy across system levels demonstrates the capability and performance of living systems.

Virtual Realism and Computation

Virtual realism is governed by the principles of computation. Virtual realism refers to the computational principles such as levels of analysis (Marr, 1982). Across levels of analysis, computation consists of problem solving across three levels – at the level of computation, the level of algorithm, and the level of physical implementation. At the level of computation, functional tasks are decomposed into parts or subcomponent tasks. The performance of specific functional tasks is considered as a hierarchical organization of subcomponent tasks. At the level of algorithm, a formal procedure provides the specification for the production of an accurate output, given a specific input. At the level of physical implementation, a device using a particular technology implements the computational principles.

The computational level of virtual realism describes the specific problems that virtual reality technology can solve, and the hierarchical organization of functional tasks and task subcomponents for problem solving. The algorithmic level of virtual realism details the specific formal procedures for the production of accurate output, given a specific input. Rule-based algorithms define specific formal procedures for the production of output given particular input.

The physical implementation of virtual reality technology implements the performance of specific functional tasks and their component parts. Virtual reality is the production of virtual environments using virtual reality technology. Virtual reality technology consists of computer hardware and software. Virtual reality hardware consists of computers with the capability for user design and testing of products in virtual environments. Virtual reality applications consist of software programs for the simulation of different social and physical capabilities in real time. Virtual reality applications are beneficial for the simulation of a range of interactions in the social and physical world through virtual technology. Virtual reality applications build knowledge systems of the social and physical environment through the design of different products in the virtual environment.

The mental life of virtual reality demonstrates properties that are consistent with complex adaptive systems. Mental life in the virtual world observes multiple levels of organization. The mental content in the virtual world consists of mental states and mental events that are of relevance across economic, social, and cultural system levels to the political context. The mental life of the virtual world demonstrates open boundaries that are malleable to the flow of information in matter and energy. The structure and organization of the mental content of the virtual world are malleability to change. The mental life of the virtual world consists of the mental states that comprise the mental content of economic, social, and cultural system levels of the political context.

The production of mental content in the virtual world is consistent with rule sets. The virtual action or virtual behavior of agents is governed by rule sets or the valuation of rules or settings of the system level. The mental content of agents in human systems is governed by laws and the cultural system level. The valuation of the rule set as control parameters is malleable and can be regulated to change the structure and organization of mental content.

The mental content of the virtual world is adaptive to system level changes. The mental content of agents and the rule sets within the system levels are adaptive to changes across time. Changes at the system level affect the environment and the structure and organization of the system level. For instance, changes at the system level of the virtual world affect the virtual environment and the structure and organization of the system level in the virtual world.

Mental life in the virtual world is consistent with self-organization. Mental life in the virtual world demonstrates patterns within component parts of the system. The self-organization of mental life in the virtual world illustrates the capability for self-governance or governance based on rule sets determined by rules of internal interaction and external constraints.

The mental content in the virtual world builds from emergent behavior at the systems level. Emergent behavior describes the interaction of agents and system components across time. The emergent behavior of interaction in virtual environments contributes to the capability and performance of system levels of societal structure and social organization.

The complexity of adaptive systems as in the virtual world demonstrates a nonlinear causality. The mental content of virtual worlds is consistent with nonlinear causality. The mental content in virtual environments is malleable to multiple feedback loops, external constraints, and initial conditions. The mental content within system levels of the virtual world is changeable and responsive to emergent property. The mental content in virtual environments may be considered emergent virtual property or the emergent property of virtual environments.

Culture and Virtual Realism

Cultural variation refers to the wide variation in the cultural beliefs, traits, and practices of shared meaning systems across groups of people defined by ancestry, language, customs, geographic origin, and ethnic heritage. Cultural selection describes the processes that guide the maintenance and replication of information across people. Cultural inheritance refers to the transmission of cultural traits across related and nonrelated individuals. Cultural evolutionary processes guide the selection of cultural traits through processes of adaptation. Cultural evolutionary processes describe the selective pressures that guide the maintenance of cultural traits.

Cultural transmission of social information through social learning demonstrates the transmission of knowledge across individuals. Cultural transmission through social learning is cumulative and malleable to the social context. Cultural transmission refers to the transmission or replication of information from the cultural model to social learners. The persistence of information from cultural model to social learner demonstrates cultural transmission.

Cultural evolutionary processes show mutual influence of environmental and cultural systems. Cultural traits are adaptive and responsive to environmental pressures. Cultural adaptations serve a protection function from environmental pressures. Cultural traits that are adaptive are maintained and transmitted across individuals in the social group. Cultural traits that demonstrate specialized functions and promote survival and reproduction may undergo further selection through dual inheritance processes of coevolution. Culture processes refer to the hierarchical structure of knowledge systems that interact at the environmental and individual levels.

Cultural computation pertains to the representation of cultural traits across levels of analysis. Cultural computation refers to the production of cultural mental content that plays a functional role in the environment. Cultural computation is the set of functional tasks and their task components that perform functions of protection at the cultural and individual level. Cultural computation consists of the functional task components for protection from environmental threats. Cultural mental computation refers to the cultural mental content that protects from environmental threats. Cultural mental computation consists of the functional tasks that perform the function of protection. Cultural mental computation refers to the mental content of cultural variation. Cultural mental content consists of the mental property of cultural variation.

Cultural mental content is physically realized across multiple environments. From physical to virtual environments, cultural mental content consists of the mental states that comprise the stream of experience (**Table 12.2**). Cultural

TABLE 12.2 Culture and virtual realism

Possible world	Physical	Digital	Virtual
Environment	Three-dimensional	Digital	Immersive
Cultural	Cultural	Digital media	Virtual culture
Social interaction	Face to face	Digital model	Agent
Social representation	Corporeal body	Digital image	Avatar
Social group	Social group	Digital group	Virtual group
Social identity	Physical identity	Digital identity	Virtual identity

mental content refers to the cultural traits that are important for cultural and psychological adaptation. Cultural mental content refers to the flow of information generated through the mental computation of culture.

Cultural transmission through social learning refers to the cultural acquisition of social knowledge from the social environment. Cultural transmission in the social environment assumes social experience and social interaction in real time in the real world. Social learning in the physical environment relies on the social processes of imitation. During social imitation, social knowledge is acquired in the social learner through the observation and imitation of a cultural model.

Cultural transmission through social learning suggests the importance of cultural transmission across possible worlds. Cultural transmission in the physical world encompasses the social learning that occurs in real time through face-to-face interaction. Cultural mental content with a functional causal role represents informational content for cultural transmission in the virtual world. Cultural transmission in the virtual world expands knowledge generation into virtual environments with distinct capabilities.

Cultural transmission in virtual environments assumes virtual experience and virtual interaction in the virtual world. Social learning in virtual environments is guided through virtual interaction with cultural avatars and agents. Humans as avatars acquire social knowledge through virtual experiences with cultural agents in the virtual environment. In immersive virtual environments, humans can control the spatial and temporal parameters of the social learning experience, including the virtual cultural identity of the cultural model and learners and the level of cultural transmission across cultural models and learners.

Immersive virtual environments facilitate social representation and social interaction in novel ways. Immersive virtual environments allow for the social representations of cultural avatars and agents through virtual replicability. Virtual replicability enables cultural avatars and agents to perform social functions through multiple virtual social representations based on a virtual identity. A virtual identity can simultaneously perform multiple social functions across possible worlds through the virtual replicability of social representation. A virtual identity can simultaneously interact across multiple social environments through the virtual replicability of social representation. From physical to virtual worlds, social representation in social interaction can occur as a continuum of social identity.

The virtual replicability of social interaction demonstrates virtual social multitasking through the virtual identity. In physical environments, the physical identity of the individual is preserved and maintained as a unitary, fixed, and stable entity. In virtual environments, the virtual identity of the individual is replicated and multitasked to perform multiple social functions in the virtual social environment. The virtual identity of the individual is realized as a complex, malleable, and dynamic entity. Virtual replicability introduces virtual properties to the notion of identity. The virtual replicability of identity suggests the augmentation of the spatial and temporal properties of mental life.

Immersive virtual environments allow for avatars to perform virtual interactions through novel virtual identities. Avatars can experience virtual interaction as a virtual identity that is distinct from physical identity. Avatars can perform virtual interactions as multiple virtual identities that are distinct from the physical identity. Avatars can control multiple virtual identities in the virtual world, while maintaining the identical physical and social identity in the real world.

Human avatars as social learners and agents as cultural models comprise virtual social networks in immersive virtual environments. As a virtual social network, avatars and agents generate novel knowledge that links groups and sustains connections. Virtual social networks coordinate the participation and engagement of groups. Virtual social networks rely on technology to enable virtual network activities. Virtual networks coordinate the cooperation of members and encourage the participation of members in network activities.

Immersive virtual environments present novel formats of mental content for cultural transmission. Virtual mental content refers to mental property that is acquired through virtual experience. Virtual experiential states of knowledge comprehension and production are beneficial as accurate representations in the structure of the world. Virtual experiential knowledge states as mental representations generated through mental and machine computation are transmittable and reflect an innovation of immersive virtual environments. The mental content of virtual knowledge states shows the valuation of accurate representations in the world. Virtual knowledge states as physical and mental states are valuable as accurate representations of experience in the natural world. The simulation of a physical and mental state in the virtual world can hold truth value as a representation of experience in the real world.

Virtual Realism in Culture

The virtual realization of cultural property, including the cultural mental content and cultural causation of mental states, illustrates a physical realizer of the cultural mind. Cultural mental life is composed of the cultural mental content that constitutes the cultural traditions and cultural identity that define cultural community. Cultural mental life consists of the mental content with causal power for the advancement of cultural participation and cultural life in the world.

Cultural computation plays an important role in the production of culture. Cultural mental content refers to the generation of cultural knowledge from mental computation. Cultural mental computation describes the mechanisms that generate mental representations of culture in biological machines. Cultural machine computation refers to the cultural mental content from computing machines.

The study of cultural computation refers to the conceptualization of the cultural system as a computational system. In computational terms, the cultural system is a physical system with mappings of the physical states of the cultural system and transitions of the physical states as operations or representations of the system. Cultural processes consist of mappings of physical states and their transitions in the cultural system. Cultural processes are operations of mappings of physical states and their transitions in the cultural system. Cultural computation assumes that the computational level describes the cultural dimension of natural phenomena in the world.

Cultural machine computation consists of the mental content produced from computing machines. Social interactions of cultural agents and cultural avatars produce mental content that is characteristic of cultural processes. The social interactions of agents and avatars produce cultural property that performs a functional causal role in the environment. The cultural computation of simple to complex machines produces the states of cultural knowledge that are readily accessible at the computational level. The cultural computation of machines demonstrates the functional causality of cultural knowledge through computation.

Mental causation in cultural life refers to the causal power of computation in the cultural sphere. Cultural computation connotes the causal roles of cultural mental content from cultural life. Participation in cultural life generates cultural mental content in the causal network of mental life that advances cultural development. Cultural life generates the mental content of protection and empowerment in cultural communities. Cultural life contributes to the shaping of societal perception that advances societal goals.

Technology is beneficial for the design and simulation of cultural systems. Cultural models contribute to the shaping of societal perception for cultural development and cultural advancement. Virtual environments support the design and development of cultural models that perform specific functional tasks. The social behavior of cultural agents and avatars contributes to the mental causation of cultural life. Cultural models generate and share mental content that contributes to the development of social knowledge. Cultural models communicate informational content that supports societal equality and participation in the cultural sphere.

Virtual environments are a useful application of technology in the cultural sphere. Virtual environments promote the design and testing of cultural products.

Virtual environments can be designed for the simulation of social interactions across cultural contexts. The simulation of social interactions across cultural contexts assumes culture-based social knowledge as a basis of mental state understanding and social cooperation. The efficacy of virtual environments to produce cultural mental content demonstrates the causal power of computation in the cultural sphere.

The design of immersive cultural environments for the training of intercultural communication is beneficial to cultural development. Virtual environments provide a platform for the design of immersive cultural environments. Immersive cultural environments allow human users to develop human avatars that acquire cultural knowledge, perform cultural learning, and demonstrate cultural competence. Immersive cultural environments may facilitate cultural training for psychological and cultural adaptation.

The computational level is necessary and sufficient for knowledge generation in cultural systems. The cultural computation of minds and machines demonstrates the functional performance and causal power of cultural systems through multiple physical realizers. Cultural mental computation generates and stores cultural knowledge through the functional architecture of the mind and brain. Cultural machine computation simulates the production of cultural knowledge through cultural computing machines.

The cultural system as a computational system situates cultural mental representations within multiple physical realizers. The causal power of the cultural system for the production of cultural change stems from the flow of information in matter and energy at the computational level. The causal role of cultural mental content to produce cultural change arises from computational dynamics of cultural life. The causal mental content of cultural systems consists of physical states and operations or representations of physical state transitions that are malleable to cultural change. The cultural system as a computational system advances the notion that the physical states of cultural systems are governed by computational principles and the laws of nature.

Culture in Virtual Realism

Cultural dimensions are foundational to the social structure of virtual worlds. Cultural computation as a simulation of the cultural system in the virtual world consists of the social characteristics that are foundational to societal organization and social norms. Cultural dimensions that define distinct self representations describe a range of social representations for social simulation with virtual characters. Virtual interaction with avatars and agents builds on the concept of self and identity.

In virtual worlds, avatars and agents act as social representations with social characteristics consistent with the concept of the self and identity. Avatars and agents display virtual behaviors consistent with their virtual identities in immersive environments. The psychovirtual identity hypothesis asserts that there exists a relational set of mental states to physical states of virtual environments. Every mental event is related to a physical state of the computational system. Psychovirtual identities describe the functional mapping of the set of relations of mental states to physical states in virtual environments.

Cultural dimensions contribute to the social simulation in virtual interaction. Cultural dimensions provide social rule sets for constructing the social representations of virtual identity. Virtual interactions simulate the physical and social characteristics of social interactions. Virtual interactions comprise the social simulation of perspective taking through social roles. Virtual social roles impart knowledge of social norms that are specific to the fulfillment of specific social responsibilities. Perspective taking through virtual interactions enhances the experiential knowledge of first- and third-person perspectives and the second-person perspective of their interaction through virtual identities.

The cultural properties of virtual identities contribute to the intergroup processes of cultural groups. Virtual identity theory (VIT) refers to the simulation of social characteristics that define identity and its membership of social groups in virtual environments. VIT refers to the physical states in the computational system that are causal-relational to virtual identity and its membership of social groups. As social identity theory (SIT) refers to the intergroup processes of social identity that define membership to ingroups and outgroups in the physical world (Tajfel, 1981), VIT focuses on the simulation of virtual identity in the virtual world and the intergroup processes that are constructed at the level of computation.

VIT posits that the intergroup processes of cultural groups through virtual identities are malleable and self-organized at the cultural system level. According to VIT, the simulation of virtual identities demonstrates the cultural malleability of identity to novel cultural rule sets. The virtual replicability of identity in virtual environments further illustrates the complexity and malleability of intergroup processes based on virtual identities. The virtual replicability of identity demonstrates how virtual identities can perform multiple social representations across social categories.

Virtual identities consist of the set of mental and physical states in the virtual world that are characteristic of specific cultural content. The cultural computation that defines virtual identities is the set of physical states in the computational system that comprise the simulation of the cultural content of the virtual identity. The set of physical states and their transitions in the computational system are sufficient for the production of virtual identities at the cultural level system. More generally, the computational system as a physical system is causal-relational to the mind at the cultural level system.

The functional mapping of sets of cultural mental states to physical states of the cultural system describes the causal-relational function. The dynamical models of cultural information flow describe the cultural patterns in the physical states of computational systems that are causal-relational to cultural mental content. Cultural information flow in mental content concerns the mental states that are malleable to cultural change. Information flow at the cultural system level depicts the physical states of matter and energy and their transitions that comprise cultural mental content.

Virtual Realism in Culture and Technology

Virtual reality technology demonstrates the capability of virtual social interaction among cultural agents and cultural groups. Cultural agents, such as computer-controlled cultural avatars, perform functional tasks that are consistent with the social roles of cultural models. Cultural agents can be effective as cultural models for social learning. Cultural agents can enhance the cultural similarity of learners to the cultural model for the acquisition of cultural knowledge through social learning. Cultural similarity of models and learners is beneficial to the cultural performance of simple and complex movements. Cultural agents are a medium of model-based social learning. Cultural agents facilitate social communication among cultural group members. Cultural agents build the social patterns of behavior that benefit from social identity and intergroup processes. Cultural agents promote content biases in social learning.

Virtual realism represents a stance of the construction of a virtual reality from the interaction of humans and computers in the virtual world. Virtual realism places the reality of social interaction among humans and computers in the virtual world. Virtual environments provide platforms for human avatars and computer agents to perform social interactions and virtual behaviors. Virtual social interactions of avatars and agents are guided by the simulation of mental state understanding. Virtual social environments construct a virtual social scene for avatars and agents to demonstrate social knowledge and perform social behaviors in the virtual world.

Virtual reality is a platform of social interaction with the capability for the social training of cultural competence. Virtual environments designed to simulate cultural contact facilitate the training of cultural skills and the building of cultural competence through the virtual interaction with cultural agents. Virtual environments that are designed to train cultural competence may consist of software programs that test levels of cultural knowledge, including language, customs, ethnic heritage, among others.

The use of virtual reality for the training of cultural competence builds on the capability of virtual environments to produce the mental content and behavior of individuals that demonstrate cultural adaptation. Virtual reality technology as a software training program may facilitate social learning of cultural learning algorithms from cultural models.

Conclusion

Virtual realism is a philosophical inquiry into the nature of mental life across possible worlds. Virtual realism explores the components of mental life and the role of computation in the natural world. Virtual realism presents novel positions to contemporary perspectives of scientific realism. The traditional view of scientific realism places emphasis on the reality of the structure of the world that is independent of the mind and the reality in the world that is a part of the mind with a causal role.

Virtual realism is a refinement of the notion of the reality of the world at the level of computation. The computational level represents a component part of the structure of the world; the computation of minds and machines are component parts with a causal role. Virtual realism posits that computation is a part of the reality of the world with a causal role. The computation of minds and machines are a part of the complex causal network of mental life. Computation consists of the component parts of reality in the world that are inherent to the structure of the world.

Virtual realism contributes to the production of culture through the use of technology for cultural production. Virtual realism as a technology demonstrates the causal interaction of culture in minds and machines. The cultural production of the mind interacts with the cultural production of the machine. The interchangeable flow of information from matter and energy demonstrates the malleability of cultural production from the computation of mind and machine. Virtual realism is a kind of physical realization of mental content. Virtual realism explores the reality of virtual environments that are a part of the structure of the world.

References

Godfrey-Smith, P. (2003). *Theory and reality: An introduction to the philosophy of science.* London: University of Chicago Press.

Marr, D. (1982). *Vision.* New York: Freeman.

Minas, H. (2014). Human security, complexity and mental health system development. In Patel, V., Minas, H., Cohen, A., Prince, M.J. (Eds.). *Global mental health: Principles and practices.* New York: Oxford University Press.

Tajfel, H. (1981). *Human groups and social categories.* Cambridge: Cambridge University Press.

Further reading

Blascovich, J. & Bailenson, J. (2012). *Infinite reality: the hidden blueprint of our virtual lives.* New York: William Morrow.

Mesoudi, A. (2009). How cultural evolutionary theory can inform social psychology and vice versa. *Psychological Review, 116(4),* 929–952.

CONCLUSION

Philosophy of Computational Cultural Neuroscience represents a comprehensive introduction to the philosophical foundations of computational cultural neuroscience. Computational cultural neuroscience is a field of study that investigates the computational principles of the cultural mind, brain, and behavior. Philosophical topics in computational cultural neuroscience explore the foundations of the nature of the mind and the philosophical implications of the principles of computation as a scientific field of study. The exploration into the computational foundations of the nature of the culture and the mind contributes to classic questions in philosophy of mind, including the relation of the mind and the world, the functional role of the mind, the mind as a causal power, and the mind as a physical instantiation.

Philosophical topics in computational cultural neuroscience introduce novel concepts into themes and topics in philosophy of mind and philosophy of science. Themes in computational cultural neuroscience address computational approaches to the study of the cultural brain, the role of computation in the cultural brain as a discovery process, and the physical realizations of the cultural mind in the biological organism and its computer simulation. Themes in computational cultural neuroscience represent an advancement of theory and methods in computational science to test causal models of cultural neural networks and cultural nets. Through the design of computational approaches for the simulation of the cultural mind, computational models contribute to the demonstration of the biological plausibility of the cultural brain. Computational approaches in cultural neuroscience highlight the use of computational discovery tools for the formal testing of cultural models.

Traditional stances in philosophy of science posit the role of the scientific process in the causation and explanation of the structure of the world. Topics in philosophy of science articulate the significance of the scientific paradigm for the use of scientific observation in theory confirmation. Contemporary philosophy of science advances inquiry into the role of science and scientific progress in naturalism and the humanistic understanding of the natural world. Contemporary approaches in philosophy of science grapple with the notion of probabilistic theory as an arrangement of possibilities in the structure of the world. From natural laws to principles of reason, philosophy of science encompasses the conceptual knowledge and explanatory rationale that govern the natural world.

Themes in computational cultural neuroscience constitute an advancement of the theory and methods in computational science for use in theory confirmation of cultural processes in the mind and brain. Computational science as a scientific paradigm consists of the principles of computation for the simulation and construction of computational models. Computational science as a contemporary scientific and technological progression is a continuity of the scientific realism in empiricism into the simulation and construction of naturalistic phenomena. Whereas naturalism represents the traditional stance of the scientific paradigm for the purpose of discovery of natural laws, computational science is the application of computational approaches for the purpose of the simulation and reconstruction of natural phenomena.

Computational science places a complementary emphasis on the philosophical positions of artificialism. Artificialism elaborates on the connection between technology and philosophy of science. Technology as a tool for scientific discovery broadens the role of computational discovery to inform core considerations in philosophy of mind and philosophy of science. Computational discovery underscores the importance of unbiased observation in computational discovery. Computational discovery suggests that the application of technological tools for scientific observation complements the role of theory-neutral observation of the scientist. Computational discovery suggests boundaries in the role of technological tools for theory-neutral observation. Computational discovery helps to inform the determination of computational value judgments.

Computational discovery implies the pertinence of the computing mind in philosophy of mind. Computational science in philosophy of mind introduces a range of notions of the computing mind and its role in the structure of the world. The role of the computing mind as a cultural mind entails the intentionality of intelligent design in the performance and production of mental computation. The elaboration of the understanding of the mind as a cultural mind is apparent from the processes of computational discovery. Varieties of artificialism address the purpose and role of computation in philosophical inquiry.

Philosophical notions in computational cultural neuroscience consider the impact of computational theory for computational discovery and its implications for scientific observation and explanation. The computing mind as a cultural computer contributes to the notion that cultural patterns of thought arise from the neural information-processing mechanisms of the biological machine. The computing mind as a cultural mind implies that mental computation performs functions at the cultural level. The cultural processes of the mind as a physical instantiation in the brain broaden the functional role of the biological machine for cultural production. Computational theory introduces novel concepts and methods for the computational discovery of a range of classic questions in philosophy of science, including culture as computation, culture as a dynamical process, the role of scientific and technological progression for the advancement of cultural processes.

In three parts, the book introduces themes and topics of philosophical inquiry into core considerations of computational cultural neuroscience. The first and second parts of the book examine fundamental themes in computational cultural neuroscience and their implications for philosophy of mind. The third part of the book explores core concepts in computational cultural neuroscience and their contribution to philosophy of science.

In Part I, the first four chapters explore the philosophical implications of concepts in computational cultural neuroscience for understanding the mind. The mental capacity to detect agency or the intention of others as a dimension of the mind is a core component of cultural and mental life (Chapter 1). This chapter explores the notion of agency and the importance of understanding intentionality in cultural and mental life. The concept of supernatural agency is the broadening of the understanding of intention into the cultural sphere. Supernatural agents represent the lawlike expression of religious or spiritual life in the cultural sphere. The simulation and construction of supernatural agents in possible worlds demonstrate a role of the concept of supernatural agency in computation at the cultural level.

Automatism as the automatic processing of information constitutes a computational component of the cultural mind (Chapter 2). The simulation and construction of cultural automatons demonstrate the automaticity of action. The cultural automaton is part of the cultural property in the superenvironment that contributes to automated production. The notion of automatism as a computational component in the mind and in the world is discussed.

Interface theory (Chapter 3) discusses the notion of the interaction of minds and machines. Interface theory considers the interaction of mental and machine computation as a point of interconnection that is deterministic and bounded. Interface theory explores the continuum of causal relations in minds and machines at the cultural level, including the commonality and incommensurability of mental and machine computation.

Machine functionalism (Chapter 4) refers to the functional role of the mind and its properties. Machine functionalism considers the functional role of mental property across multiple physical realizers. Machine functionalism underscores the causal power of the mind as a part of the physical system. The chapter explores the notion of the causal-functional role of mental property at the cultural level.

In Part II, four chapters explore themes in computational cultural neuroscience and their philosophical implications for the mind. Reconstructionism

(Chapter 5) as a complementary approach to reductionism refers to the lower level of information processing. Reconstructionist notions place emphasis on the emergent property in the physical world as the property from the interaction of parts. Emergent property contributes to the parts and particulars of the total causal structure of the world. Reconstructionist notions consider the importance of emergent property in the cultural mind and as part of the cultural level of the organized system.

Machine physicalism (Chapter 6) highlights the physical parts and particulars of mental and machine computation in the physical world. Machine physicalism expands the notion of machine functionalism or the functional role of machine computation for the mind into the realm of spatiotemporal dimensions in the physical world. Machine physicalism entails the spatiotemporal properties of machine computation at the cultural level in the physical world.

Computational theory of mind (Chapter 7) discusses the mental and physical property of mental state understanding. Computational theory of mind consists of the mental and physical states that comprise mental state inferences at the level of computation. Mental state inference at the computational level is deterministic, produced from a particular input–output relation. Computational theory of mind contributes to the mental performance of social inference at the cultural level.

Simulation (Chapter 8) refers to the understanding of mental states. Simulation as the modeling of mental state inference in the world consists of multilevel mechanisms. Mental state understanding refers to the real-world properties of experience at the cultural level. Simulation contributes to the understanding of mental states from the information processing of neurobiological mechanisms.

In Part III, the last four chapters discuss fundamental concepts of computation at the cultural level and their implications for philosophy of science. The last part of the book discusses foundational concepts in computation and their role in the structure of science.

Artificialism (Chapter 9) explores the computation of minds and machines and its implications for philosophy of mind and philosophy of science. Artificialism as a philosophical stance considers the impact of computation on artificialistic approaches to science. Artificialistic approaches to science consider the use of computational discovery for information production. The chapter elaborates on the impact of machine computation on culture.

Machine learning (Chapter 10) discusses the role of computational discovery for scientific processes of explanation. Machine learning refers to mental performance for explanatory inference. Machine learning algorithms detect patterns and regularities in the structure of the world. The knowledge extraction from large datasets in the world contributes to the patterns and regularities that inform prediction. The chapter discusses the importance of machine learning for the cultural level of the organized system.

Intelligence (Chapter 11) considers the notion that intelligence is a part of the mentality and thought of minds and machines. Intelligence and its multiple forms contribute to information production as parts of the organized system. Intelligence is a foundational concept of mental and machine computation. Intelligence defines a standard of criteria in the functional performance of minds and machines. The intelligence of minds and machines is the mental content that defines mentality. Intelligence as a part of possible worlds demonstrates the functional role of computation.

Virtual realism (Chapter 12) explores the notion that machine computation is a source of knowledge of reality of the world. Virtual realism as a part of scientific realism develops the role of computation in minds and machines in the maintenance of production of knowledge and reality across possible worlds. From the physical to virtual worlds, computation contributes to the simulation and construction of experience and reality.

There are several novel concepts from computational cultural neuroscience elaborated in the themes and topics of philosophy of mind. Cultural processes of mental and machine computation comprise the parts and particulars that contribute to the cultural level of the physical system in the physical world. The notion of the cultural mind consists of multiple physical realizations. Understanding the causal-functional role of the mind contributes to fundamental philosophical inquiry into the nature of the mind across possible worlds.

The book also introduces several themes in computational cultural neuroscience into core considerations in philosophy of science. The language and concepts from computation place emphasis on complementary approaches to traditional notions in philosophy of science. Artificialism as a complementary approach to naturalism is one of the concepts that are introduced into core considerations in philosophy of science. Naturalistic approaches to the understanding of the natural phenomena in the world are standards in the structure of science and the scientific tradition. The introduction of machine computation into the production of mental content that has a foundational role in the scientific tradition opens a range of issues that require elaboration in the scientific tradition. The contemporary reliance on computation for discovery processes implies the necessity for a philosophical position that is distinguished from naturalism.

The development and implementation of programs of research in computational cultural neuroscience provide evidence-based knowledge of the computational principles underlying cultural processes in the structure and function of the nervous system. The contemporary notions of the cultural mind as a cultural computing machine present novel consideration for philosophical inquiry. The use of computational discovery in scientific thought introduces concepts and language into the scientific tradition that further advance core considerations.

Scientific advancement in computational cultural neuroscience contributes to levels of cultural development and the advancement of cultural life of individuals, societies, and nations. The use of evidence-based knowledge in computational cultural neuroscience for the development of applications in cultural computation provides novel tools and approaches for cultural development. Scientific and technological approaches further intellectual and societal commitment to cultural advancement and contribute to the promotion of the quality of cultural and mental life for all.

INDEX

Note: Page numbers in **bold** indicate tables.

For Product Safety Concerns and Information please contact our EU
representative GPSR@taylorandfrancis.com
Taylor & Francis Verlag GmbH, Kaufingerstraße 24, 80331 München, Germany

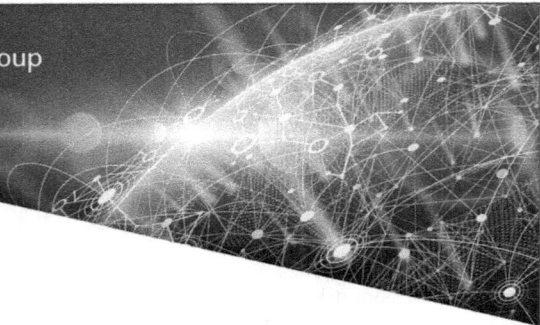